SURFACE TENSION

BY
GARY HOLLEN

PublishAmerica
Baltimore

© 2005 by Gary Hollen.

All rights reserved. No part of this book may be reproduced, stored in a retrieval system or transmitted in any form or by any means without the prior written permission of the publishers, except by a reviewer who may quote brief passages in a review to be printed in a newspaper, magazine or journal.

First printing

Micki,
Meeting you at the TATone christmas Parties and Sharing Conversations has been Special!
Gary

ISBN: 1-4137-8372-4
PUBLISHED BY PUBLISHAMERICA, LLLP
www.publishamerica.com
Baltimore

Printed in the United States of America

ACKNOWLEDGMENTS

I gratefully acknowledge those special people who offered critical suggestions and badly needed assistance. Because of them, I was inspired to write, and finally was able to breathe life into my story. Along the way I found my voice.

> Lucile D. (Mom) Bartel
> Kim Campbell
> Martha Grace Jossi
> Kim Irwin
> Teri Irish

I also gratefully acknowledge supporters who offered encouragement, patience, and understanding.

> Marri and Jason Ashley
> Mandi Hollen
> Liz Hollen

TABLE OF CONTENTS

Remembrance	7
A First Flirtation	16
A Grand Old School	34
Osaka-Wa-Wa	38
Fingerprints	72
A White Sport Coat and a Pink Carnation	93
Phi Alpha, Brotherhood, and Minerva, Too	111
Picking up the Pieces	124
Down but not Out	132
Parade Rest	140
A New Bearing	144
Commencement	154
An Unexpected Encounter	157
A Rendezvous	162
Bell-bottomed Trousers	176
Boston Baked Beans	196
Steak and Lobster	211
Scrambled Eggs	228
An Old Salt	253
Now and Forever	275
You Are My Sunshine	280
Windows	288
Releasing Fruit	305
Over the Bounding Main	324
The Spy Game	343
The Journey	359
Second Chances	366

REMEMBRANCE

The brilliant psychoanalytical beam that probed and studied the oft-times misunderstood illnesses of the human mind was extinguished in 1939. The same year, Lou Gehrig bade a somber farewell to baseball. Another light, a frightening beacon of world discord, was born not far away when Poland crumbled under the might of Nazi Germany. Onto a lesser-documented scene, a minor event occurred when the first cries of Lee Grady were heard in the maternity ward of St. Charles Hospital in Bend, Oregon.

Lee Grady and his family moved frequently. A trail littered with moving boxes originating in Bend, wound through Portland, Oregon City, and returned to Bend before finally stopping in Salem. Through it all, Lee experienced rationing of gasoline, sugar and even nylon stockings. More personal, perhaps, were the blackouts, visits by the iceman, country schools, the simple pleasure of enjoying a first stick of Wriggle's gum, and the inconvenience of an outdoor privy.

When did he first awaken to the world around him? Was it when he celebrated the aroma of his father's jacket, discarded after coming home from working a shift at the Kaiser Shipyards? Maybe it was the gamut of communicable diseases he endured or the trips he made to the eye doctor to correct a lazy eye. Then again, maybe it was experiencing a nation's sadness, wondering why his mother's cheeks were stained with tears the day the president died.

Through it all, whether he called home a city dwelling, a farm, or quarters behind a mom and pop grocery story, Lee gradually passed into his teens. Along the way, the loss of his grandfathers deeply affected him. Their gifts to him were his parents, an appreciation of the golden rule, and a love for our national pastime.

Life was simpler, pleasures less complex. A family reading circle, a trip to the drive-in theater to enjoy a picnic, nightly sessions in front of the Motorola listening to the Bronx Bomber dispose of another victim or the Shadow solving another mystery typified the times. So too were games such as marbles, sandlot baseball, playground basketball or football, boxing matches held in a neighbor's shed, nightly bouts of hide and seek or kick the can, or an afternoon fantasy of playing cowboys and Indians.

Initially an allowance earned for doing chores provided his spending money, but earning real money in the strawberry and bean fields beckoned when his family moved to Salem.

With the strawberry season at the Marsters' farm concluded, his first Salem friendship with the Marsters' son, Mel, initiated, Lee moved on to a different picking challenge. A peek-a-boo game with the long slender green beans found in the extensive foliage of the Abbott pole bean field, offered a wage of two and one-half cents per pound, a quarter cent bonus for staying the entire picking season. With a steady, always persistent effort, he earned six dollars a day. No small achievement, the sum was generated only after picking three hundred pounds of the elusive legumes.

On a particular hot muggy day in late July, the sun had just reached a position announcing the arrival of lunchtime. Sunday Meade, an attractive freshman friend of Lee's sister, had occupied the other side of his rows most noticeably throughout the morning. Trying every imaginable way to engage him in a discussion about anything, everything, she intently directed untiring efforts towards enticing him to participate in an activity having no particular relation to picking beans. His quiet appeal ignited her interest, but boredom and toiling under the hot sun prompted her to take more direct, extreme measures.

"Have any of your friends done it? You know, had sex?"

A deafening pause ensued.

"I asked if you've ever done it. Well?"

The silence persisted as he continued an assault on the heavily loaded bean vines.

I wonder why she's asking me? Like I've had so much experience!

He held no interest in girls; the feeling usually reciprocated. Most of the time he ignored them completely because the objects of his inattention unnerved him. He was resigned, perhaps content, to let his persona lie hidden behind the dark, plastic glasses he wore, a focal point of his slight, sinewy stature.

Undeterred, she continued to move up the rows with him as though being pulled along on a leash. In one final effort, she boldly offered herself to his aloof scrutiny, pausing seductively to rest on her haunches. Sighing loud enough to cause him to turn his head towards her, she leaned forward through the vines and rested alluringly on her hands. A total look of innocence was painted on her face as she peered up at him, unashamedly revealing the top of her breasts unprotected by her partially open blouse. Much to her disappointment, only guarded, darting glances overruled an apparent lack of interest in the enticing display. Undeterred, she brushed her hand slowly across her chest. Releasing another button of her blouse, she sighed even more audibly.

Surprise was quickly replaced by a shy smile quickly spreading across his face. Nodding his approval of the inviting display, he stared in awe.

Holy cow!

At once, he realized that only a scanty undergarment contained the prominent nipples capping her snow-white mounds.

"Wha, what did you say?"

She chuckled, swiftly moving to the attack.

"Have you ever done it before?"

He shook his head, nervously continuing to study the tempting display.

"No! Heavens, no! Why, have you?"

Shrugging nonchalantly; she offered, "Maybe. Um, yeah, a few times."

"How does that translate, once, twice, more?"

She grinned, confidently assuming control.

"Does it matter? Hmm, it bothers you when I tell you I'm not a virgin anymore, doesn't it?"

His cheeks reddened.

"Not really. Um, how many times have you done it, then?"

"I haven't counted. Um, several times." Shrugging nonchalantly and she confessed, "Ah, I lost my virginity a long time ago."

She paused to evaluate his reaction. Noting a heightened curiosity, she bent further forward, revealing even more.

"I don't do it with just anyone. I'm real choosy."

Suddenly nervous and uneasy, he again began to fill his pail with the captured treasure from the vines while still managing to steal periodic glimpses of the prominent mass pressing against her blouse.

"What makes you say 'yes' to one person and 'no' to another? Where do you get the courage to do it anyway? Aren't you afraid of...?"

"Getting pregnant?"

She shrugged casually. Inhaling deeply, slowly, but deliberately, she held her breath just long enough and then slowly exhaled.

"Not really. Ah, nobody makes me do anything. I do it if and when I want to. It's like, ah, I just know it's the right thing to do. You know what that's like, don't ya?"

"Ah, no. I've never felt that way before, about anyone."

He paused for a moment with his eyes still fixed on her offering. Finally he shrugged in resignation.

"I'd probably be afraid to do it, but I guess I can see how a person could come to feel that way, especially with you."

His face reddened immediately. The inexcusable revelation was just cause to compel him to again chase after the hard-to-find string beans.

Fuck! Why did I have to say that? Sunday doesn't belong in the bean fields. The clothes she wears aren't what you see the rest of the girls wearing, and she's the only one I know of who can put in a full day picking and still walk away as though she has never left home. Even on a good day of picking, she never gets dirty.

As his struggle for something to say continued, he noted Sunday's overpowering features. Without any effort, she exuded femininity through the entirety of her shapely torso. Her lips invited contact, passionate tender communication. Long dark eyelashes introduced penetrating deep blue eyes pleading to be noticed, and then gazed at. Her other physical features, years advanced from other girls he had met, featured a thin waist and flat tummy flowing into inviting, perfectly formed hips. When she inhaled, Sunday's rising mounds solicited physical communion. Nothing about her female structure, even her shapely long legs, appeared to have the slightest hint of athleticism. Even so, there was amazingly no evidence of any extra dimension anywhere on her body. She seemed to have a physical form constructed for one purpose.

An idea festering since he first noticed her suddenly interrupted his thoughts.

I wonder what it would be like to do it with her?

Noting his silent inventory, she smiled coyly.

"Wanta see more?"

He gulped, turning an immediate deep shade of red. Nervously, he fixed his eyes on the ground.

"Aren't you afraid someone will see?"

"Don't worry about the rest of them. The only thing they care about right now is going to lunch. Trust me, you're the only person who can or cares to see."

Leaning back on her haunches, with only the thick vines protecting her from his view, she invitingly thrust out her substantial chest. Inviting mounds, only partially covered by the low-cut white bra had at last captured his full attention. She reached behind her back and unsnapped the obstructing barrier, lifting it to release the treasure from bondage. Glancing at him, she smiled at the fascination frozen on his face. Mutely, she inched towards him.

"Well?"

"Wow! They're huge! Are they as soft as they look?"

"Why don't you touch them and see?"

He blushed. Suddenly the back of his neck was burning, his stomach churning.

"I don't know. Think I should?"

"Why not? Come on, it's okay. I like having the right person touch them."

"Are you sure? My hands are all dirty."

"It's okay. I'm going to take a shower when I get home, anyway." Chuckling, she teased, "Besides, I'm a very down-to-earth kinda girl."

Slowly, methodically he started to move his hands towards her as she again began inching towards him. Pausing, she reached out to take his hands, slowly guiding them towards her inviting breasts.

As his fingers contacted the white undergarment, she convincingly suggested, "Put your hands underneath and get a real good feel."

Slowly, ever so gently he obeyed her command, shuddering as he contained her huge breasts.

"Wow!"

"Squeeze 'em and run your hands all over. Umm, oh God!" She whimpered, "Why don't you touch my nipples. I really love to have my tits played with."

Fingering her soft, spongy caps, he rasped, "They're so soft. Holy shit! The nipples are starting to get hard. How am I doing, does it feel okay?"

She nodded, obviously enjoying the gentle tours of her sensitive flesh.

"Hold still, Lee. I'm gonna grab it."

Without invitation she reached between his legs and started to squeeze the expanding shielded mass.

"How's that? Is that good or what?"

She studied him, a look of innocence replaced by a devilish grin.

"Wanta go some place where we can let this guy out to play?"

He gasped, "I don't know, Sunday. It's not bad right here."

"Um, I don't want to pick anymore. Let's leave. We could go over to the creek and take a dip or something, huh?" Smiling as she gently tended to his increasing excitement, she teased, "Like it, don't ya? I can take care of you here, but it would be a lot better over there."

Studying him intently, she started making rapid movements with her hand.

"Anyway, it's too hot to pick any more."

Moving with the rhythm of her movements, he gasped, "I didn't bring any trunks."

"We'll swim in the nude, silly."

Suddenly her hand stalled. Smiling seductively, she removed her hand. Reaching down to pick up a clod of dirt, she playfully tossed it at him.

"Oh, my. Look at you. You're all hot and bothered. Want me to play with it some more?"

He nodded.

Moving his hips forward, he offered, "Sure, if you want. Um, you want me to play with 'em some more?"

She grinned, suggesting, "We don't have to go swimming, Lee."
"What would we do?"
Smiling coyly, she confessed, "I want to do it with you."
"Do what?"
She snorted.
"Shit, let's get naked and do it. I want you to fuck me. Ah, you'd like to, wouldn't you?"

Something about what she said, the bold offering of sex, completely unnerved him. He removed his hands from her breasts, brushing her hand away. For a brief moment he sat on his side of the row stunned, speechless, glaring at her nervously.

"I've never done it before."

Nodding towards the distant grove, she started to button her blouse.

"There's a soft grassy area over there under those trees. Almost nobody ever goes over there, except me." Smiling as she rose and turned to leave, she added, "I go there every day."

"Does anyone ever go over there with you?"

"Sometimes."

"I suppose you do it with them, huh?"

Shrugging, as though it was not big deal, she offered, "Sometimes, it all depends."

"On what?"

"Shit, Lee, I have to like 'em."

"Can't people see what you are doing?"

"The grass is real tall. It's hard to see anything, especially when you're lying down." Smirking, she continued, "There's a ditch over there I've made a bed in. When I lie down in it, nobody can see me even if they know where to look."

She stared intently at him, hesitating for another moment.

"Come on, let's go over there and do it. You know you wanta."

After a brief hesitation, he shook his head.

"I'm not sure. No, I don't think so, Sunday. No, not today."

He continued to study her out of the corner of his eye as he returned to the task of filling his pail. She looked so alone, so vulnerable. For an instant she had the appearance of a small child who had lost a favorite toy.

"Sunday, I'm going to pick for another few minutes and then go to lunch. I have to keep picking so I can make my quota for the day. Besides I don't want to take the chance Lloydene will find out."

"I wouldn't tell. Holy crap, Lee, I'm not crazy. Your sister and I are friends."

"Maybe you would and maybe you wouldn't. I'm still not in the mood. Maybe I'll feel different later."

"This could be your last chance, you know?"

"I'll just have to take that chance. Um, by the way, thanks for picking with me today. If I don't see you after lunch, maybe I'll see you on the bus. I like you, Sunday, but I'd like to get to know you a little better before..."

He silently studied the ground. After a moment, he looked up.

Damn, she's gone.

Slumping over the bucket, he stared for what seemed like an eternity at nothing in particular. Crouched on the soft muddy ground, he considered her offer.

Who am I kidding? She probably couldn't find anyone else to spend her noon break with.

Shaking his head, the full realization of what he probably represented to others rang a silent bell in his subconscious. Nobody had to tell him he blended into a crowd. Even his dark hair and a skin coloration that darkened easily when exposed to the sun weren't unique. Only his ruddy cheeks, easily tinted when he was embarrassed, caught people's eyes. Because of that perception, he had a tendency to hang back, choosing not to seek center stage.

Sunday's bold suggestion had caught him completely off-guard, disarmed him. Still, he wondered what it would be like to be alone with her to share her considerable assets. He felt a confused, ambivalent attraction to her. Her shapely body emitted a sensual invitation that absolutely intoxicated him. During the brief mutual exploration in their row, she had cast an exhilarating aroma of sexual desire. Still, she made him feel uneasy.

I don't know why, but Sunday kinda turns me off. It would probably be fun to.....Ah, I doubt I'd know what to do, anyway. When I get home from picking, I've got to get ready to go swimming with the Lee twins at Mill Creek. They said something about jumping off the bridge that crosses Mill Creek out near the State Pen.

Sunday introduced a complex challenge, forever altering how Lee would view the entire male-female issue. More involved than putting berries into a hallock or legumes in to a pail, the payoff had consequences far exceeding any amount of monetary compensation. Everyone has needs and a means to satisfy them. In the wink of an eye, the minute increment of time, his code of personal interaction had been transformed into something more than a simple set of rules, an agreement between participants in some meaningless sandlot game.

A FIRST FLIRTATION

When a person is ten, life is uncertain. Fear of death, yours and others', is disquieting. At sixteen, visions of life lasting forever, the twenty-first season can't come soon enough! Thirty comes and goes with the looming threat of forty, annual reminders emphasizing life's rapid passing. Fifty is the eye-opener, dulled senses and diminished flexibility. At sixty, the aching back serves as a reminder of over-optimistic physical exertions from earlier engagements and present appointments. Whoa! Slow down! It's time to stop and smell the roses, take some time to remember how it was. Forget all those other appointments. Take a moment to recall the events of those teenage years.

A joy of participation drove Lee to play games, participating in a variety of contests. The passion was fueled by the thrill derived from competition. Whether in a marble ring, risking dated marbles, or on a dimly lighted court in Bend playing horse, he was avid in his pursuit. As he advanced through the grades, sandlot contests and nightly neighborhood involvement with kick the can or hide and seek amusements gave way to more serious pursuits. Organized baseball, basketball, and football all captured his attention, attracting his energy. Consumed, he had a fantasy of playing professionally. Regrettably, he could only offer average physical size and mediocre ability.

At the time, it was doubtful he would ever overcome a discomfort of being around girls. He was reserved. Acknowledging female presence made him nervous. If a girl dared initiate even the most casual verbal exchange, his speech meter probably resembled a person with a speech impediment. Content to run with a crowd, some exceptions noted, whose major claim to fame was regular school attendance, he supported the school's athletic teams and frequented movies at the Capitol or Tower theaters. He and the majority of his friends didn't seek attention. They were comfortable operating in the shadows.

Thoughts about the fairer sex occupied some of his time. During those infrequent moments, perplexed, confused feelings probably waged a ferocious battle. On one side, a voice might have suggested he seek them out, while the other more dominating messenger most likely advised he was better off without them. Girls, sometimes a curiosity, made him feel strange. Credence to this contention resulted when he allowed his idle thoughts to dwell on a particularly attractive classmate during one of his classes. A visual investigation of her alluring torso encouraged his mind to pan to a moment of fantasy. After the brief escape, he abruptly returned to an acute state of awareness. The lapse had created an untimely anatomical reaction, the inopportune boner. Embarrassing!

After a long, nondescript summer spent toiling in the canneries, his senior year began with a thunderous bang. Like an electric storm, everything seemed to happen at once. School clothes had to be bought, careful selection a must. Necessary activities included getting reacquainted with friends, firming up plans for the final year, and giving some consideration to which college to attend. He also needed to find a part-time job to earn some badly needed spending money.

Traditionally, by default if not by decree, the seniors occupied the top rung of the school's social ladder. He and his classmates, acknowledging their newly acquired eminence, made sure their attire illuminated the new ranking. Their wardrobes soon boasted of woolen oversized overcoats, colorful sport shirts, dress slacks, and heavy brogan shoes. More casual dress included an ensemble consisting of a lettermen jacket with black sleeves and a full red body, a button-down shirt, sweater, white cords, and white bucks or brown saddle shoes. Those shoes, at first a fad, soon became fashion statements. Argyle socks were barely visible but considered a necessity.

Hairstyles varied. A small minority sported the new ducktail cut. Lee elected to stay with the conventional flattop. Shaving was merely an exercise to erase unseemly fuzz. It was also an excuse to splash ample amounts of cologne on his face. The pungent aroma might accomplish what a lack of confidence might not, render the opposite gender defenseless to a back seat move. Of course, he had to work up the nerve to ask a girl for a date, first.

Prepared to take the school by storm, like many of the people he looked up to, he had become a finely coiffured specimen.

The women covered up everything, wearing long, mid-calf skirts. Pleated creations with a Pendleton Woolen Mill label sewn inside, for all electing to snoop, the skirts were embellished with several trendy mini-slip undergarments. They tucked two, no more than four of them under their skirts to provide a revealing flare, exposing an encouraging view of the lower leg. When fully fitted for display, the bearer looked as though she had passed unsuspectingly over an updraft created by a powerful ventilator. Completing the ensemble were black flats and color-coordinated stockings marking their legs slightly above the ankle.

If not the beginning, the year spanning his senior year marked the start of a new era in publishing with the introduction of the scantily clad centerfold. However daring the pose, it still broadcast a portrait denying bold uncensored exhibitionism. It stimulated curiosity, while still managing to challenge the imagination of those most directed to fantasize about what treasure might be hidden. Unlike the publication's racy foldout, senior girls shunned teasing a viewer's fantasy. Sweaters or long-sleeved blouses, buttoned snugly up to the neck, added the finishing touches to their outfits. On game days they wore white pep-club sweaters that had, on the back, black and red Viking emblems. A badge with their names embroidered on it was placed on the sweater slightly above their left breast. Body-hugging pleated skirts completed the outfit.

The Oregon State Fair served as a catalyst for overcoming courting challenges with many a casual introduction occurring on the midway. Most North Salem girls deemed worthy of forming a relationship with for the upcoming school year were there. The brood included some underclass chicks as well as the older, upperclass hens. Lee didn't take this courting activity too seriously, as he planned to begin his quest when school began.

After the school year began, joining those who hadn't already found someone, he began to gather essential information about the available flock. The first to be scrutinized were the incoming sophomores, as new talent always received the first review. The more attractive had already been taken.

Junior and senior coeds who had not been claimed from the year before were considered next. A shy, mildly attractive senior caught his eye as he shrewdly searched the halls for a suitable object of his affection. Initial intuitive assessment confirmed she was interested, smiling whenever he looked her way. When he muttered a hurried expression of greeting, she responded in kind. Had he failed to notice he had developed into the type of young man who could attract a girl's attention?

Teenaged males rarely focused on an ideal female companion without first looking at their oblique qualities. The criterion for selecting a companion read like a short shopping list for a movie snack. Brief, it contained little if any substance. They were looking for a girl who displayed a well-constructed proportional shape, all shapely curves radiating from all of the proper places. Further, those wondrous arcs of inviting flesh had to be firmly symmetric, even to the most critical and discerning eye. Stacked with model-like facial features was essential. Finally, she should be cooperative, as a psychological or physiological block known as libido influenced most decisions. Most boys thought with an anatomical appendage that although an important part of the anatomy had no designed purpose for supporting the thought process.

Susan was above average. Although it didn't matter then, the aging process would flatter her. At the moment, however, she didn't display all of the important components deemed necessary in determining the perfect woman. Modestly camouflaged, her assets to the searching, inquisitive eye were ordinary. She wore glasses, not flattering, but, then again, they didn't detract from her qualities. She portrayed an attractiveness that captured his attention, but her charm and pleasing personality cinched the deal.

He surmised she didn't intend to improve some young stud's image, doubting she would ever allow herself to become a topic of locker room conversation. Following carefully defined steps, a script of holding hands followed by a goodnight kiss, parking, necking, petting in the front or back seat of some Detroit special was okay for others, but she had loftier goals.

All signs indicated Susan was offended by small talk. Casual discussions with her in physics class convinced him she preferred in-depth exchanges, searching for a deeper meaning. Thankfully she wasn't impatient like he was, offering a counterbalance to his sequential, thirst for solving puzzles. She conveyed a refreshing randomness.

The time had come for him to unfurl, like a flag kept in protective storage, the handsome features he had kept hiding in the shadows. He decided to ask Susan for a date.

That evening, his presence was a mere formality at dinner, as his thoughts were occupied with plotting out a game plan. Finally he excused himself from the table, having barely touched his dinner, to beat a hasty retreat to his room. He needed to rehearse the intricate script he had created. Women, so he'd been told, wouldn't respond to clumsy, unpolished efforts. He had to be cool and smooth.

Finally, at seven o'clock, an advised best time to call, he had finished all necessary preparations. Quitting his room, he headed slowly toward the phone in the kitchen. Once there, he sat down much the same as he would if being interviewed for a job for the first time. Poised in front of the black telephone, he stared for a moment at the dial. Methodically, he withdrew a piece of paper from his pocket. Her telephone number, clearly printed in black, bold numerals, loomed in front of him. Intently, he studied the number, finally lifting the receiver. Putting it to his ear, he reached toward the dial with his forefinger and methodically began dialing the seven digits of her number. After dialing the last digit, a minute nerve-racking pause ensued. Not yet ready to make contact, he frantically returned the receiver to its cradle.

Several more times he duplicated the effort, bravely dialing her number, cowardly hanging up before the connection had been made.

She won't recognize me. If she does, she'll make some excuse for not wanting to talk.

Finally, all resistance lost out to one daring impulse. Dialing the seventh and final number, he anchored his free hand to the table to make canceling the call impossible.

Maybe the phone will be busy.

No luck, the phone rang! No turning back, he changed his petition.

Maybe she's not home.

"Hello!"

Curtly forceful, the person answering made Lee shudder.

"Hello! Is anyone there?"

Gathering his thoughts, Lee attempted to muster the nerve to speak.

Finally he muttered a fragmented interrogative, "Susan, Susan Sloan, she there?"

Muffled and distorted, his inquiry clumsily invited the respondent to translate.

The voice at the other end snapped back with exasperation etched in his voice, "What did you say? You'll have to speak clearer, son."

Apparently the person belonging to strange voice didn't care how his victim felt or about the increasing state of anxiety he experienced. If he did, he displayed unrelenting determination to keep his prey uncomfortable, dangling helplessly until tiring of toying with him. What a fine fisherman he must have been. Finally losing interest in continuing to play the wriggling prey, Mr. Sloan cut him loose and laid down the receiver.

"Susan! It must be for you. Maybe you'll have better luck finding out what the caller wants."

Melodically Susan's voice intoned, "Hello? This is Susan speaking."

Words stumbled from his mouth, gained stability, and, then finally, momentum.

"Hi. Um, this is Lee Grady." Grasping for a life rope, he continued, "How did you do on the physics lab today?"

With the initial foray completed, a feeling of relief unraveled within like a soothing afternoon breeze on a hot summer day as he waited for her reply.

"Oh, hi. How are you? I'm fine, thanks. I think the lab went okay. Ah, I had a little trouble understanding how to figure the mechanical advantage of pulleys. I'll bet you didn't, though, huh?"

"Not too much."

Thoughts about the next topic to introduce swirled about in his head. He didn't bother to notice the long pause, content to celebrate her apparent happiness in receiving his call. Finally he broke the spell.

"Ah, if you ever need some assistance, I'd be glad to help."

"Thank you. I'll keep it in mind. Knowing me and math, I'll probably be tapping you on the shoulder sooner than you might expect."

Another uncomfortably long pause paralyzed the phone lines, freed only by her kindness.

"I think it's going to be a good game Friday, don't you?"

"Oh, yeah. That's why I called. Ah, I wanted to know if, ah, would like to go with me to the game, on Friday."

"Oh, yes, I'd love to go! I'm so glad you called."

"You are? Great! Ah, well then, I'll see you tomorrow at school. We can discuss the plans then."

He really had no idea what they talked about. At Susan's end, if she had even considered fending off his overture, it was denied by the swift frontal assault he launched. Avoiding any prospect for a quick end run, he targeted her jugular. More realistically, she probably agreed to go to the game as a protection against the backwash of another long, awkward pause in the conversation.

Wow, she said yes.

He pumped his fists into the air.

Yes! Yes! Damn, was I cool or what? Oops! Oh, God. I forgot to tell her what time I would pick her up. I wonder if she wants to double date?

He shook his head in frustration.

Hmm, I gotta figure a way to meet up with her so it appears natural, and, ah, there's no way I want her to find out that I think going out with her is a real big deal. I gotta let her know it's cool, but…I wonder what she's gonna think when I tell her I forgot what time I'm picking her up?

The next day at school, he faced the unnerving task of confirming and arranging the terms of their Friday night engagement. First he had to find her. By the time the last bell of the day sounded, they somehow failed to run into each other than a brief acknowledgment during physics class. After putting his books away in his locker, he decided to go look for her. Without looking, he wheeled around.

"What the…?"

"Oomph!"

Susan grunted as she took a spiraling dive towards the hard landing strip of the hallway floor.

He reached out in an attempt to catch her. Losing his balance, he tumbled forward onto the slick waxed floor beside her.

"Crap! Look what I've done now." Scrambling to his knees, he grunted, "Are you okay?"

"I think so. Whew, you really pack a wallop."

"I was hoping to run into you today, but I didn't plan on meeting like this."

"What was wrong with seeing me during physics class? It might have saved me from this."

"Ah, I had other things to do. You sure you're all right? I hope I didn't hurt you."

A crowd of curious students started to gather about the accident scene as she started to make an attempt to adjust her nearly prone position to a more dignified sprawl. Subconsciously adjusting her tilted glasses, she made a perplexed, smoothing swipes at her mussed hair.

"You know, Lee, with all of these books, papers and stuff from my purse spread all over, don't you think it would make an interesting collage?"

"No kidding."

The closer Lee looked at the frustration she displayed, the funnier the situation appeared.

"I'm sorry for laughing, Susan, but you should see how you look."

Recognizing the humor of the situation, she started to snicker. The surrounding crowd also began to laugh. Midst the commotion, she masterfully fulfilled her role as the producer and director of the upcoming engagement.

"I've been trying to make a connection with you all day, but the way you ignored me in physics, I wasn't sure you still wanted to go to the game on Friday."

"I'm sorry. I've had a lot on my mind."

"Apology accepted. Um, if you would like to pick me up at seven, it'll give us plenty of time to get to the game before kickoff time. How does that sound?"

"Seven's good."

"I hope you don't want to double."

"Absolutely not. Ah, double dating is such a hassle."

Operating with the expertise of a ringmaster, she continued to orchestrate.

"I hope we're going to the dance. I love to dance."

"Absolutely. I really like to dance, too."

"You said something about looking for me?"

In less time than it would take to explain an unexcused absence, Susan, in a most professional, adept manner, massaged his ego with flattery, reconstructing the plans to her liking. Enlightened, he had received a first lesson, a crash course.

"I did, but you seem to have just about covered everything."

I actually said all that last night? Wow! I was cooler than I thought.

Friday night, his parents handed him with the keys to the family chariot, a 1954 four-door Nash Rambler. The cherry Detroit special, a stick-shift with the shift located on the steering column, was a faded pale green with well-worn reclining seats hidden by a color-coordinated covering. The car did little to stir automotive passion. It was transportation.

Three minutes before seven, he pulled up in front of Susan's home, carefully navigating the automotive masterpiece into a parking spot near the walkway leading to the front door. Her home, located on Fifteenth Street, was only three blocks from the North Salem campus. Simple in design, it was a white, single leveled house with a basement. The front porch light, accenting her home's yellow trim, was lighted.

I wonder if I should shake Mr. Sloan's hand? Um, I'll let him make the first move. What about Mrs. Sloan, I wonder what I should do with her? That's it, I'll just smile and say hello or something. From the sound of his voice, I'll bet Susan's dad is gigantic. He'll probably threaten me with great physical pain or her mother will cast a spell upon me if I let anything happen to the apple of their eye tonight. I just hope I fit the image they'd allow to go out their daughter.

The door opened before the last refrain of the doorbell's attack on the musical scale had concluded.

"Hi. Let me get the screen for you."

He grinned, nodding as she started to struggle with the screen.

"The doorbell, ah, music to announce you guests?"

"Darn, this stupid thing never opens when you want it to."

She continued to push with bullheaded determination as he stepped forward to survey the problem.

"The doorbell, I hate it."

Chuckling, he suggested, "Why don't you unhook the screen?"

"Oh yeah. Um, could we start all over?"

"Sure. By the way, I hate your doorbell, too."

"You don't like the tune?"

"My tastes are a little simpler. A simple buzz would do."

Smiling as she motioned for him to enter, she suggested, "This way to the living room. I'd like to introduce you to my parents before we leave. Um, don't say anything about the doorbell. I think they kinda like it."

Susan led him to the living room where her parents were seated on the couch.

"Mom and Dad, I want you to meet Lee Grady."

Mr. Sloan rose, extending his hand.

"I'm pleased to meet you, Lee. Honey, this is the young man I grilled the other night."

"Be nice, Bill." Smiling warmly, Mrs. Sloan continued, "I'm pleased to meet you, Lee. I don't like the doorbell, either. Why don't you make yourself comfortable? If you're not in a rush, we can talk a little before you and Susan leave."

Lee had barely made an indentation in the beige divan when Susan's mother opened the bidding. "Have you always gone to North Salem?"

"Um, I moved to Salem from Bend two years ago." His face flushed, adding, "I just met Susan. We have physics together."

"Mom, Lee's just being modest. He played on the junior varsity football team last year. Ah, we were also in English class together."

"I like people who don't brag about their accomplishments."

"It's what my wife likes most about me. What do you plan to do after you graduate, Lee?"

"I don't know for sure, but I'm thinking seriously about going to college."

Mrs. Sloan nodding her approval, acclaimed, "Good. Do you know where, yet?"

"Mom, I think we should leave. You and Lee could probably talk all night, but between you and me, I don't want to miss the kickoff."

"Mrs. Sloan, I've boiled it down to OSC and Willamette. Ah, it's been a pleasure meeting you folks. I want to thank you for allowing your daughter to go to the game with me this evening. Maybe we can finish our conversation some other time. Oh, yeah, I'll have your daughter home by midnight if that's okay?"

"We enjoyed meeting you, too, Lee. I'd like very much to finish our little chat. Um, midnight is fine."

Mr. Sloan walked over to Susan and gave her a hug.

"You two have a good time. Drive safe."

Finally free from introductions and polite inquiries, they approached his car, its faded green paint job appearing much brighter under the dim street light.

"It isn't much, but it should get us to the game."

He opened the door for her, watching intently as she started to guide her body onto the decorative seat.

"Your parents are real nice."

"Thanks. Ah, we have a Rambler, too."

"Don't you love 'em?"

"My parents? Of course! Um, I'm really glad you liked them. They get a little pushy, but at least I know they care. They liked you, too."

"I meant, don't you just love a Nash?"

"Sure. We've had a lot of fun at the drive-in theater picnicking. We even got to see the movie."

"Our family, too. I guess you haven't lived until you've eaten an entire meal at an outdoor movie."

He closed the door and quickly made his way around the car as she inched towards his side of the seat.

Hmm, at least she doesn't think of me as a chauffeur. She's moving over so she can sit close. That's a good sign. Yeah, things are looking good. I'm sure glad I asked her out.

"I love football. What about you?"

"Basketball's my favorite." Grinning, he probed, "I suppose you watch football on television with your dad?"

"As a matter of fact, I do. What's more, we even discuss strategy."

"I can understand guys liking to watch football, but..." Frowning, he offered, "You're not like most girls. They don't understand football let alone like it. They probably go to games just to socialize."

"You're probably right." She squeezed his arm as she confessed, "I'm glad you asked me to go to the game with you. Oh, look. There's the Leslie Junior High School parking."

The brick structure attached to South High was the north end of an almost perfectly shaped rectangular horseshoe with the football field enclosed inside.

"There's a parking spot over there near the entrance to the field. Hurry, I think we can get to it before the other car does."

I wonder what everybody's going to think about me taking Susan to the game? I sure hope everyone thinks I made a good choice.

As they started the short march towards the grandstands, the previous glow inexplicably disappeared. He was suddenly uncomfortable and ill at ease about bringing her.

I could always let her walk in front of me. That way, nobody would know we're here together. Yeah, I could make some excuse about it being crowded or something.

He had little time to explore thought. Grabbing his arm, she nudged him in the direction of the bleachers.

"Let's sit up there. We'll have a better view of the field."

In a matter of less than a second they had noticeably become a couple.

Quickly they settled onto the hard, narrow bleacher seats near the top of the North student section. Someone nearby, elbowing his girlfriend, mumbled audibly, "Will you look at whom Susan came with. Isn't that Lee Grady?"

His companion nodded, turning towards them.

"Susan, you didn't tell me you and Lee were seeing each other."

"This is our first date."

The National Anthem saved them from further interrogation. After the Vikings received the kickoff, they settled into an almost mute observance of the action on the field. Occasionally commenting on a call by an official or evaluating a play, they were contented to remain in their private worlds.

Man, watching this game reminds me of what it was like to play football.

He shook his head in disgust.

Now, that's something I'd just as soon forget. The only reason I turned out my junior year was because I got cut from the basketball team as a sophomore and I wanted to be a part of one of the teams. Great choice! Each night was pure hell. It was kinda like a bloodletting exercise. I was just scrimmage fodder for the bigger guys on the team. At least when I forced my tired, bruised body through the front door of my home, my parents showed that they cared.

He grinned, suppressing a more audible chuckle.

I don't know how many times they begged me to quit. Ah, I guess I was stubborn, but they did tell me to always finish what I started.

He leaned back onto an empty bleacher seat behind him, clasping his hands behind his head.

I never did appreciate the chemistry class I took. There was something about the odor emitted by the experiments had to conduct that totally turned me off. Even so, it was kinda like a sanctuary, a place to mentally prepare myself for the beating was gonna receive at football practice. Yeah, after school, my locker took over. Nope, I wouldn't play football again if it were the last thing on earth to do.

"The second half is about to start."

"Huh? Oh yeah. I guess I was daydreaming."

She smiled.

"Really? Well, welcome back. Ah, to bring you up to date, the balls on our forty-yard line, second and two to go. We've got a nice drive going, so what do you think we should run? Personally, I favor a quick look-in pass to the end. It's been open all evening."

"Why not go long? It's a free play."

"Hmm, that could work, too."

True to her suggestion, the next play was a slant to the end, resulting in a twenty-yard gain.

"How did you know they'd run that play?"

"Does it bother you I enjoy football and know a little about it?"

He put his arm around her and squeezed affectionately.

"Not at all. I think it's cool. Um, I'm having a good time. How about you?"

"Me, too, but with the beating we're taking, I'm starting to look forward to the dance."

"You mean trailing only forty to nothing, we don't still have a chance?"

"Are you kidding? This game was over at halftime. Ah, let's go. I can't take any more of this. If we head for the car now, we'll beat the rush."

"Good idea. The dance is apt to be crowded. Maybe we can get there before it's totally jammed." Chuckling, he inquired, "Did you wear padding?"

The drive across town to the Viking campus was spent eulogizing the team's loss. Susan's considerable knowledge of football continued to impress him, displaying an uncanny ability to review, with insight, the results of the North debacle.

At a stop light near Meier and Franks, he inquired, "What do you think about going to Up-Town Drive-in after the dance?"

"Sounds good if we have enough time. The dance doesn't get over until eleven."

"We can play it by ear. We may not even be in the mood by then. Well, what do you know? We've arrived."

"Hmm, we made good time. Usually it takes longer to drive over here from South Salem."

"Yeah. Hey, we're in luck. There's an empty parking place."

"It may be the only one left. Suppose someone left early?"

"Could be. Ah, I'll bet it's crowded in there."

An overwhelming entanglement of gyrating humanity greeted them inside the outside gym. Guarding against the crush of the crowd, he slipped his arm around her and pulled her towards him. Responding to his gesture, she slipped her arm around his waist.

"I doubt we'll get any closer to the middle of the dance floor than we are right now. Wanta dance?"

"Sure. Suppose they call this operating on the fringes?"

Suddenly choreographed movements capturing the provocative mood of sultry lyrics and an energetic rock-and-roll beat erupted. Adept bopsters began twisting their bodies into the convoluted movements suggested by the new musical beat.

"There's no way."

"We could try doing the swing."

"If we don't get run over." Shrugging resignation, she surrendered, "Why not! It's not as suggestive and reckless as what they're doing. You'd think with all that erratic movement, they'd dislocate their back or something."

"You mean you don't wanta scoot around the floor with your feet spread apart and your upper body parallel to the floor?"

"Be serious. This style of dancing is much more fun. I'd never able to do that new, whatever it's called, dance. I don't know how they do it. If it was me doing all that twisting, I'd be in traction for a month."

Several forty-fives later, fatigue setting in, Lee wiped his brow.

"I need a break! Would you like to sit out one? I'm dripping wet."

"Me, too. If it's okay, I'm gonna touch base with some friends I promised to look up. Man, I need to cool off."

He nodded as another Presley song began to broadcast its message to the assemblage.

"Go ahead. It'll give me a chance to find out how the Lee twins feel the game went."

Grinning, she offered, "Badly. See you in a few."

"No kidding. Oh, I see 'em over there with all the guys on the football team."

It doesn't matter what the situation is. Susan's able to handle about everything.

When he reached the Lee twins, he turned to see if she was okay.

Hmm, she's something else. I'm sure glad I asked her out.

The Lee twins, ends on the football team, were still wound up tighter than a drum, carrying on excitedly about the game.

"I can't believe the catch I made on the slant pass Mike threw to me."

"The one he threw to you before Mark caught the down and out?"

"Yeah, I had to go up for it in a huge crowd of Corvallis defenders."

"It was a great catch, Marty. You really put on some moves after you caught the ball. You looked like a six-foot five-inch tailback with all the shiftiness of a wild pig. That was a nice catch you made, too, Mark."

He glanced in Susan's direction.

I thought she was looking up some friends. Hmm, I wonder who he is?

Marty grinned, subtly pointing in the direction where Susan was located.

"Oh, oh. Somebody's movin' in on your date."

"I see him!"

"Don't get upset. I just wanted to make sure you knew what was going on."

"Sorry. Hmm, I wonder where he's from?"

"He's the starting fullback for the Saxons."

"Didn't they have a game tonight?"

Mark interrupted, "They play tomorrow night. When we play a home game they either play on Saturday or go out of town to play, remember? Oops, speaking of Susan..."

She tapped Lee on the shoulder, grinning faintly.

"Lee, Paul asked me to dance with him. Do you mind?"

"Go ahead. I'll stay here and talk to Mark and Marty. We still have some things to discuss about the game."

She squeezed his arm, smiling gratefully.

"I'll be back in a few."

Mark chuckled, jabbing Lee playfully in the ribs.

"Lee? Is it okay if I dance with Paul?"

Lee's face flushed. The loud music interrupted before he could offer any objection to the obvious barb.

"So, are you going to start taking her out?"

"I don't know, Marty." Scanning the floor, he revealed, "I'm thinking about it. Ah, it's a little early to tell. I do kinda like her, though."

"I take my hat off to you, fella. You haven't been at this long, but you've already got them asking for permission to dance. I'm impressed!"

"I doubt she was asking permission. She was just being polite."

"Not to worry, I hear he has hot pants for Susan's best friend."

"Really!"

A strikingly attractive girl suddenly appeared in his field of vision. Responding to acknowledgments that were thrown her way as she passed through the mob.

Holy cow, there's Mary Anne Giard. She's one the most popular girls in school. I'll bet she's in nearly every school activity in addition to being the student body secretary. Untouchable! I think there must be a sign somewhere with blinking lights warning everyone to keep their hands off. Anyway, she only dates college guys. All the time she's sat next to me in English class, she's never said more than an occasional 'Hi, how are ya?'"

Mark noted her approach, nudging him to get his attention.

"Do you see who's coming this way?"

"Yeah, wonder which football player she's looking for?"

"Well, I know she's not coming to see me." Grinning, he teased, "You're all on your own, Lee. I think you're the one she wants to see."

She probably wants to talk with one of the football players. Besides, what would she want to talk to me about, anyway? We hardly know each other.

Dodging in and around the many animate objects blocking her path, she continued to close the distance between them.

What if she really does want to talk to me? Nah, who am I kidding? She wouldn't waste the time.

She stopped directly in front of him. Surveying the appreciative throng of onlookers, she coyly bowed.

"Hi there. I don't think I've seen you at one of these dances, before." Smiling, she continued, "Thank you, Lee. I'd love to dance. Um, on second thought, maybe we should wait for next song."

He blushed, her forward demeanor overpowering him.

"Ah, yeah, this one's a little fast for me."

Seconds later, the hand grenades and bomb explosions ended.

Winking, she melodically suggested, "Shall we try it now?"

Responding to her suggestion, he took her in his arms. Swept into a vortex of an unexplainable setting, he sensed every eye fixed upon him as he visualized hundreds of eyes boring into the back of his neck.

She's a good dancer, so light on her feet.

He didn't dare look at her. At that moment, a sudden fear his weaker eye would not cooperate encouraged him not to make direct eye contact.

She showed a complete contentment to follow his every move with defined grace and a fine tuned sense of anticipation. Marianne seemed to flow with the music, his every suggestion. Repaying his unspoken complement as the music ended, she smiled as his eyes made darting contact.

"Darn, the music ended too soon. You're a very good dancer, Lee."

The peaceful, melodic sound of a Johnny Mathis ballad gave way to the violent pulsating music of another Presley tune.

"Yeah! I think that's my cue to return to the folks I came here with tonight. Give me a call, sometime."

Just as she came, she left, blending into the mass hysteria of the pulsating mob. Gazing at the place where she had stood scant seconds before, he was unaffected by the interruption of a familiar voice.

"Was that Mary Anne Giard you were dancing with a moment ago?"

"I'm fine."

"Huh? Hmm, not what I asked, but I'm glad to hear it."

"Oh, hi, Susan. I must have been daydreaming. Did you enjoy dancing with Paul?"

He turned again to see if he could see Mary Anne.

"Where does Paul go to school?"

His first significant date ended as Susan's curfew expired. On the well-lighted front porch of her home, he fidgeted nervously.

"I really had a good time tonight."

She smiled as she leaned forward and kissed him on the cheek.

"I'd like to see you again. It doesn't just have to be in class, you know."

His face reddened.

Wow, my first kiss.

"Um, I'll call you."

He wasn't completely at ease in the presence of girls, but he longer wanted to operate in the shadows, either. No longer would he work overtime to avoid the former dreaded object of his inattention.

A GRAND OLD SCHOOL

Can a chance meeting, too brief an intimacy, followed by a hasty parting, ever represent anything more than a blip on life's continuum? Will the brief liaison flower then die, or will it go into hibernation, then spring to life after a period of dormancy?

Salem High School's campus occupied an area of nearly foursquare blocks. On it, at the intersection of Fourteenth and D Streets, rested a two-story brick masterpiece with a large pillared entrance. Its entry, majestic, usually caused an awakening of reverence, as, on a miniature scale, it evoked memories of a Roman concrete esplanade leading to the building where senators awaited the triumphant return of their Emperor. Seemingly impervious to the erosion of time, its architectural design spanned decades.

The old school had a rich history steeped in tradition. Memories, too many to count, were made amidst special events and occurrences, but in the fall of 1954 it all changed when Salem added a second high school. Salem High School was renamed North Salem High, keeping the Viking school nickname, and the new school, South Salem High, became the Saxons.

Many factors shape the direction a person takes, likewise, a school's course after its inaugural ceremony. It's been said it takes a village to raise a child. In like manner, a school flourishes because of community support or special members of its student body. Nothing, however, is more important than its faculty.

North's principal, Mr. Clarkeston, a distinguished leader, commanded everyone's respect because of a unquenched love for learning, holding those charged with imparting their knowledge and experiences onto others in highest esteem. He also loved the students under his charge, possessing an uncanny ability to remember each and every name. He wasn't flashy. He simply motivated everyone to do his or her best. Presiding over a talented faculty, he fanned the flame of learning by encouraging members of the North student body to excel at something, anything.

Many on the North faculty played important roles. One of them, a physical education instructor, prowled the grounds like an old bear. Seemingly everywhere, as a baseball coach, teacher, or role model, Mr. Tanaska defied description. George, no disrespect intended, affected students as nobody else could. A hulk of a man with a deep gravely voice, he had a grip of steel. If a student didn't experience his vise-like grip on their shoulder while he affectionately inquired how they were doing or what was going on in their life, they must have slept through a three-year stay at North High. He often boasted, "I never met a bad kid."

A civics teacher, Miss Bayre, awakened deeper considerations of the troubling ideas pervading America's landscape. Her classroom windows appropriately looked out onto the front grounds of the school. Next door to the library, it occupied a position to the left of the impressive entry. Her learning chamber's location seemed to match her objective of encouraging all students to access the future through an exploration of some of life's major issues, as within those walls she challenged students to think and test their ideas. Communism, the plight of the American Negro, and seeking truths from distortions created by a propagandized view were common topics Lee took interest in while attempting to avoid a spray she emitted when her excitement erupted during a lesson.

Mr. DuBoir, the dean of students, patrolled the halls like the stealth bomber of today. Head cocked and ears perked up to detect the slightest transgression, he seemed to always nab the perpetrator before the act was consummated. He was there when students needed his counsel the most, and like so many on the North faculty, he was so much more than a teacher. He was an inspiration.

By Lee's senior year, the rivalry between North and South had split

Salem's loyalties in half, not too dissimilar from the North-South division during our country's Civil War days. The Vikings had floundered during the first two years of the split, accomplishments not easy to come by. Winning on the gridiron was rare; results in track and field, better. While the wrestlers grappled with moderate success, the basketball team was only able to show a glimmer of hope for the future. Thankfully, the baseball team saved the day.

Repeated victories by the cross-town rival Saxons had caused a huge sore to develop and fester, the wounded pride of the Viking faithful extremely tender. The class of '57 yearned for the opportunity to send the cocky Saxons down to defeat, success in football on hold for another year. Nothing repulsed the Vikings more than suffering a beating from the upstart Saxons. Certainly pride in a school is more than a reflection of athletic success, but winning on the athletic field serves as a catalyst, a means of building school pride.

As educational institutions evolve and refine their ability to educate the whole being, or as approaching manhood is formally celebrated in other cultures, Lee forecast an approach to maturity when he stopped using a tidy hall locker for assurance. Suddenly more comfortable prowling the halls with an expanding number of friends, he enjoyed pulling pranks or looking for the attention of a favorite coed.

Likewise, North's fortunes evolved. Growth and change is sometimes accomplished through a painful, difficult journey. Along the way, a variety of frustrations can be experienced, but encountering success is so sweet. Such was the Viking's victory when the basketball team ended nearly a three-year drought and secured bragging rights to the city.

North had one of the largest lineups in the state's proud basketball history, several members showing an ability to play at the college level. For some reason, though, they started the season slowly. With their bigger, slower players struggling to overcome the quicker style employed by opposing teams, they floundered. They couldn't seem to find a rhythm, but experienced observers didn't panic. Believing it was only a matter of time before the Viking's level of play matured, they prophesied, "When the coach finds the right combination of players to carry out his game plan, they'll start winning."

At the mid-point of the season, the Viking record stood at unimpressive seven wins and five losses. Suddenly answering the restless anticipation of their supporters, they exploded with a timely flurry to salvage the season. When they defeated the Saxons, they also earned a trip to the state basketball tournament in Eugene. Not realized since the glory days of old Salem High School, the student body and community actually started to dream of winning a state championship. North Salem was tournament bound!

OSAKA-WA-WA

During those moments when you escape the rigor of every day life to ponder unsettled events of the day, eyes closed in reflection; you gradually drift into the protected recesses of a deep, never-ending chasm called memory. It takes courage to release the tapes for review, the brief retreat summoning a long forgotten experience. Perhaps one of those events gathered headlines, but more probable, it didn't create enough interest to occupy more than some obscure corner of the back page of the local newspaper. Such was how the class of '57 entered the final quarter of their senior year. Songs were sung to celebrate it and newspaper articles chronicled it, but it was the Vikings who achieved it.

Mel called a few days before the start of the tournament.
"If you're interested, I've got a place for us to stay in Eugene. Sure would be better than driving back and forth to see the games."
"I can't afford a motel."
"What if I told you we can stay with my aunt and uncle?"
"Hmm. What would it cost us?"
"Nothing other than what we choose to spend on entertainment."
"I'm all for it, but shouldn't we at least buy them a gift?"
"Great idea. I'll get my mom to pick up something for us."
"Let me know what it costs so I can pay my share. Um, how far away from McArthur Court is their place?"

"A couple of miles. Hey, I was thinking, ah, we don't play until Wednesday afternoon, but maybe we should go down Tuesday to get settled in. What do you think?"

"It's okay with me. Um, are we gonna have a curfew?"

"I think we can pretty much come and go as we please. My aunt and uncle are pretty cool about those kinda things, but we'll still have to use some common sense. I think dragging in way past midnight all the time wouldn't be too good of an idea."

"This is sounding better all the time."

"You might want to have your mom give mine a call to find out about the accommodations. You know mothers."

"Yeah, and mine won't be satisfied until she know every detail."

The mid-Willamette Valley college town was a perfect place for the tournament. Everything from motels to restaurants, grocery outlets, laundry facilities and more were centrally located within walking distance of the campus. A relief to parents, all tournament activities were to take place at Mac Court or the Erb, so nobody had reason to leave the campus until it was time to retire to their motels for the evening.

The U of O had gone to great lengths to insure their guests were made to feel welcome. With the student body on spring break, university officials had, in effect, turned the campus over to the high school invaders. Accepting the university's open invitation, sixteen teams and their supporters, after landing on campus, claimed the Erb as their hangout. A convenient place to get snack, or even breakfast and lunch, it was also where dances were held after the games had concluded.

Beautiful spring weather, a week of independence, and an opportunity to make new acquaintances suddenly turned a quiet, furloughed campus into a beehive of activity. Opportunities to view exciting basketball games and additional bonuses too numerous to itemize awaited the energetic, hormonal-driven teenagers. A notable example was the line of classic, flame-painted cars parked in front of the Erb for the hot rod enthusiast's to drool over. Everyone milling about the display on wheels could feel the electricity of atmosphere surging around the row of "California Specials."

In Wednesday afternoon's opening session, the Vikings prevailed in a close game against a scrappy team from Pendleton. North's coach summarized the game in an article appearing in the *Oregon Journal's* sport's page the next morning.

"Things didn't look too good for us until Mark, ah, Mark Lee came off the bench to light a fire under the rest of my boys." As the article continued, the North coach addressed his team's tentative play. "Maybe it was first-round jitters. I don't know. Whatever it was, I can't say enough about our opponent. Not having to face Buckaroos again is a real blessing. They were well coached and played as hard as any team we've played this year."

Mark hadn't played the role many felt he deserved, his brother contributing even less, but he was an offensive machine fully capable of scoring volumes of points in a short span of time. Although Coach Paldano preferred to go with a more physical, defensive-minded lineup, Mark's contribution in a substitute role was very welcome.

After the last first round evening game had concluded, huge numbers of students from all the tournament schools gathered in upper ballroom of the Erb Memorial Student Center for the dance. Mel and Lee had planned to attend an off-campus party with several other North supporters, but answered the call by the rally squad to join a packed upstairs ballroom to show school spirit. Viewing the assemblage, Lee's enthusiasm bubbled over.

"Damn, Mel, will you look at all the good-looking girls running around this place? Suppose we could delay going to the party for a while?"

"I was thinking the same thing. Suddenly I have an urge to do a little dancing. Bingo! Do you see the foxy brunette wearing the Lincoln High School pep sweater standing over there by the stairs?"

"I sure do. Ah, what are you waiting for?"

"I'm trying to figure out if she's alone."

"It doesn't look like it to me, but from what I can see, you better make a move before someone else beats your time. She's a fox."

"What about you?"

"I'll probably look up Mary Anne and see if...ah, then again, finding her in this mob might prove difficult if not impossible."

"You might have better luck if you looked for someone else."

Lee nodded towards the stairs.

"Go on, get out of here. Don't worry about me, if I don't hook up with her, I'll look around and see what I can find." He quickly surveyed the large enclosure, hesitated, and turned back towards his friend, proposing, "Ah, you might mind find out if the gal from Lincoln has a friend."

"I'll see what I can do, but first I have to find out if she's even interested in me. She may have a boyfriend around here somewhere."

Out of the corner of his eye, Lee spotted a large gathering of Vikings. Mary Anne, as usual, was the center of attention.

What the hell, I've got nothing to lose. I'll go over and see what she has planned for the evening.

The door slammed loudly.

"Sorry, Lee. I'm with Greg. He's staying on campus so we can spend some time together. Call me sometime."

Lee nodded. The tall, muscular behemoth wearing a University of Oregon jacket was stiff competition.

Crap, I should never have quit on Susan. Chasing after Mary Anne has been a complete waste of time. Hmm, maybe going to the party isn't such a bad idea after all. Then again, seeing at all the fantastic women roaming around this place, you'd think there would be someone available. Maybe I should drift on back to see how Mel's doing.

Mel intercepted him as he neared the place where they had split up to pursue companionship for the evening.

"Hey, Lee, I've been looking all over for you. I want you to come with me."

"What's up?"

"There's someone I think you're gonna want to meet."

"Did you connect with the Lincoln gal?"

"I'm working on it. Come on, I'm gonna introduce you to her sister."

"Not the one standing next to–"

Ignoring Lee's inquiry, Mel proceeded, "Ah, Lee, I'd like you to meet Jennifer James. Jennifer, this is my friend, Lee Grady."

Smiling cordially, Lee replied, "I'm pleased to meet you, Jennifer."

He couldn't take his eyes off an exquisite beauty standing near Jennifer.

She's not wearing Lincoln's colors. Hmm, I wonder who she is? Damn, she's good looking. Nah, not even close, she's absolutely beautiful. Man, would I love to meet her.

"Likewise. Um, I'd like for you to meet my sister, Sandy James."

"Excuse me. What'd you say? I guess I wasn't listening."

"I said, this is Sandy, ah, my sister."

Is she kidding? I must be dreaming. Suppose she'd like to dance? He stared at her, completely entranced.

What a fantastic figure! What am I saying? Everything about her is perfect.

As their eyes met eyes, her soft blue orbs, like magnificent jewels, insured the trance he had fallen prey to would last. Like magnets, they enticed him to maintain contact. On this occasion he wouldn't self-consciously look away, fearful his eye would turn in.

"You have beautiful eyes."

Leaning forward, she whispered, "Thank you. So do you."

"And you say your name is Sandy."

"Sorry. My sister's guilty of that disclosure. Hmm, it's not the noise. Could it be you're suffering from a memory loss? Maybe you have trouble concentrating? I hope it's not serious because I'm not medically trained."

Blushing, he clarified. "No, my mind was just wandering. And, of course, you must be from Lincoln?"

"If I were, wouldn't I be wearing Cardinal colors?"

"Yeah, I suppose. Well, now we've established you're not a Cardinal, what are you?"

"Does it matter?"

"I suppose not."

She can't be Jennifer's twin, can she? No, that couldn't be. She seems to be so much older, so mature. Hmm, if Jennifer was a junior or sophomore, yeah, that has to be it. No way that she's a student here at the University. Whatever, appears to be very polished, worldly. Ah, to hell with it. What difference does it make?

"If you're through giving me the once-over, I think you and I should get away from the love birds so they can have a chance to get to know each other a little better. I'd like to dance."

"So, if you're not from Lincoln, ah, what...? You're not a student here at the U of O, are you?"

"Good guess!" Smiling, her eyes twinkling, she disclosed, "Course it isn't like you had too many options. The U of O is the only university I know of that's located in Eugene."

"You're kidding me! You're a student at the university? Ah, what's your major?"

"No, I'm not kidding and I'm majoring in English and political science. Um, after I graduate this June, I'm going to travel for awhile. Eventually, I plan to go on to law school. Criminal law absolutely fascinates me."

"Hmm, move over Clarence Darrow."

"I'm impressed. I see you're familiar with one of my heroes. So, what are your thoughts about our age difference? Does it bother you?"

"I don't know, I haven't had time to really think about it, but, ah, as long as you're gentle, I won't complain too much."

"You could pose a problem. Well anyway, what does four or five years have to do with sharing a dance and having a little innocent fun? Age is so overrated." Chuckling, she explained, "It's up to you to do your job keeping me occupied so Mel and Jennifer will have a chance to really get acquainted. Jen would never leave me alone in this den of libelous teenagers unless she was sure I was being entertained properly."

"Ah ha, it looks like it's up to me to keep an old lady from dying of loneliness. I guess caring for the elderly does have its rewards. Ah, it might help to know a few specifics."

"Uh-huh, and what might they be?"

"Oh, I don't know. How about telling me what you're doing here?"

"Shhh, don't say another word. I'd like to listen to what your eyes have to say. They're so expressive."

"What?"

"You have bedroom eyes."

"Would that be good or bad?"

"It could be either. Your eyes, they're so restful and reassuring. Ah, did you know they've captured my attention?"

"Like yours did to me?"

Blushing, she continued, "They're also very seductive. It's like they're offering an invitation to come to your bedroom. Shame on you! You've already laid back the covers."

Blushing, he protested, "What are you talking about? I haven't–"

"At a moment like this, maybe it's best not to say anything."

"Probably. Um, do you have a boyfriend?"

"Oh, ho, aren't you the sneaky one? Why should it matter to you, anyway? We're just going to enjoy a dance or two, aren't we?"

"I'm sorry. I didn't mean to pry."

"Apology accepted. Ah, what makes you think I have a boyfriend?"

Wrinkling up his nose, he disclosed, "Someone as good-looking as you are should have guys falling all over themselves."

"Now you're causing me to blush. You know? You're a real load, dark and handsome. Uh-huh, you're a little more than I expected."

"Not tall, though."

"Just right."

"Now I don't know what to say."

"Then don't say anything. Um, I'm not seeing anyone right now. I have much too much left to accomplish. Ah, it's something like not being able to afford allowing unimportant things like getting involved derail my locomotive. Um, I'll bet you don't have very many free moments, yourself."

"Maybe I should pinch myself. Oh crap, why'd I have to say that?"

"A Freudian slip, perhaps?"

"A what?"

Chuckling, she teased, "You're not into inflicting pain on yourself, are you?"

"No, it's just, ah, what I meant to say didn't come out right."

"You could start over. Try telling me what's really on your mind."

"What I really want to say was how fortunate I feel to have met you. You're absolutely breathtaking. I mean, you're really beautiful."

"Well, thank you again. My, you're so full of compliments and far too kind. You also make too many assumptions."

"I do?"

"Uh-huh. What if I told you I was available for the rest of the week?" Winking, she proposed, "I could be yours to pamper and spoil, if you like."

"Really? But, ah, we hardly know each other. Why would you want to spend the rest of the week with someone…?"

"Like you?"

Her steel blue eyes focused on him like a riveting gun ready to connect two steel beams.

"I'm serious. You captured my attention the moment we were introduced. Does it make sense to say I've felt really strange ever since?" Whispering, she continued, "A good kind of strange."

He croaked, "You make me feel kinda weird inside, too. Um, I vote yes. I'd love to spend some time with you."

"What do you mean, some? I did say the rest of the week, didn't I?"

"I, ah, I didn't think you meant the entire week."

"What have you got to lose? We're just going to hang out at here, dance a little, and get to know each other. Sounds pretty harmless to me."

Nodding, he concurred, "And we do have to give Mel and Jennifer an opportunity to connect, right?"

"Uh-huh. Who knows? We might even make a connection."

Taking her hands, he added, "They're hitting it off pretty well."

"We're not doing so badly ourselves."

"Would you like to meet me here tomorrow night after the game?"

"I was afraid you wouldn't ask. There's nothing I would rather do."

"Maybe we should dance. Standing here in the middle of the dance floor is a little conspicuous."

"Good idea. Um, Lee, dancing close is good. Ah, it makes it so much easier to talk."

She grinned as he slowly embraced her.

"Question?"

"Answer."

"Do your cheeks always turn red when you're embarrassed?"

"I'm a little nervous."

"Why?"

"Maybe I'm afraid you'll lose interest."

Snuggling closer, she revealed, "I don't waste time with uninspiring people."

"Me either. Would you mind if I ask you something?"

"Shoot."

"How old are you, anyway?"

"You weren't listening."

"You didn't say anything about how old you were, did you?"

"I told you I was graduating this year." Grinning, she challenged, "Now I suppose you're going to tell me age isn't important, huh?"

"Not really. I need the information for the date book you asked me about. It would be nice if I started filling it up."

"Like it isn't already full. Um, I recently turned twenty-one." Studying him carefully, she probed, "Can you handle it?"

"Maybe I should be asking you the same question. College girls don't usually spend time with someone who's only seventeen years old."

"Touché! Maybe I've learned age isn't such a big deal. Of course, immaturity, not matter how old a person is, makes a difference. Really, Lee, does our age difference matter to you?"

"How am I supposed to know? So far, it doesn't seem to bother me a bit. I kinda like older women. They're so much more…"

"Sophisticated?" Grinning, she offered, "Believe it or not, you have a quiet sophistication. Right now, all you lack is confidence."

"Really?"

"You'd be surprised at some of the children I've met since coming to school here. I've gone out with seniors who act like they're still in preschool. Ah, they also have certain fixations."

"How so?"

"They like to play with toys. I don't enjoy being someone's toy."

On the way back to their living quarters, Mel and Lee rode in silence.

Shattering the silence, Mel offered, "I really want to thank you for helping me out. It gave me the chance to get to know Jennifer a lot better."

"Not a problem, but maybe I'm the one who should be thanking you." He grinned, subtly inquiring, "So things went well with Jennifer?"

"Oh, yeah. It looks like we're going to be spending quite a bit of time together. You don't seem to be doing too bad, yourself."

"I wouldn't want to break up the sister act."

"I hope what you just said means you have plans to see her again. Ah, how old is she, anyway?"

"She's a senior here at the U of O?"

"You're kidding!" Grinning, he teased, "And she picked you? She's so damned good-looking. Ah, what do you think about keeping them occupied for the rest of the tournament?"

"It's okay with me. Sandy mentioned something about hanging out for the rest of the week, but I'm not gonna hold my breath. After she starts thinking about spending time with a high school kid, she'll probably get cold feet and cancel out."

"Oh, I don't know. I wouldn't sell yourself short."

"We're supposed to meet up tomorrow night. Wanta make it a foursome?"

Mel turned into his aunt and uncle's driveway.

"It's okay by me. Jennifer was hoping you and Sandy would hit it off."

"Oh, yeah?"

"Yeah. Ah, she wasn't so sure it would work out, though."

"Did she say why?"

"I guess Sandy's really particular. She doesn't just go out with anybody." He smiled, suggesting, "Man, this could turn out to be one hell of a week."

The next morning, in the student center, Mel and Lee studied the *Oregon Journal's* account of the tournament's first round.

"Wow! Here's a writer smitten by our rally squad's 'exuberant personalities.'"

"Really? What else does he have to say?"

"Here, read it for yourself."

"Would you mind reading it to me? I'm busy looking over the scoring summaries of the games played yesterday. Wow! Dennis is second in individual scoring for the tournament."

"Sure. Ah, he writes, 'The North Salem Rally Squad captured the crowd's fancy during Wednesday's first round of play at the state basketball tournament in Eugene. Performing their dance routines to various musical renditions, they absolutely mesmerized the Mac Court crowd. This reporter had difficulty determining where the North Salem throng began or ended, as a growing number of supporters appeared to be jumping aboard the Viking bandwagon. Every time the high-stepping lassies performed their routines, their support appeared to have grown. North Salem's rally squad stands at the head of this year's pep squads, if not all squads I've observed for the past ten years.' Quite an article, huh?"

"You're right, he does seem to have a thing for them."

"There's more. He goes on to say, 'Today when North Salem plays the North Bend squad from Coos Bay, they will need much more than some high stepping lassies to get the job done. If they stumble coming out of the blocks like they did against Pendleton, they'll be in for a long evening. Even Osaka-Wa-Wa won't save them.' Hmm, a think our girls impressed him?"

"Why not? They've earned it."

"Yeah, and when our band starts playing and the crowd start singing our fight song, everything just…"

"I know. With our rally squad out front dancing and the crowd going nuts, it's pretty impressive." Leaping to his feet, Mel blurt out excitedly, "Now you've got me all fired up."

Voicing the words and clapping his hands, a few amused students from other schools started to gather. Before long, some of the Vikings faithful had joined them, turning the relative quiet of the dining area into an impromptu pep rally.

OSAKA WA-WA
WHISKY WEE-WEE
ORE-GO-NI-A
RAH FOR SALEM HI
WHEN WE ROUGH-HOUSE
NORTH BEND HIGH
WE WILL HOLLER
BULLA-BULLA: RAH-RAH-RAH

In an afternoon contest, the Vikings again took care of business, dispatching North Bend to the consolation bracket. Again, Marty Lee came off the bench to supply the necessary spark. In an evening contest, Lincoln overwhelmed a scrappy team from Grants Pass, setting up a showdown with the Vikings the next evening.

Jennifer and Sandy's unmerciful heckling dampened the elation of the Viking victory.

Jennifer, unusually full of herself, boasted, "I don't have a doubt in my mind we're going to win. It's just a question of how much. What do you think, Sandy? Think it'll be over by half-time?"

"I'd stake my reputation on it, Jen. The Vikings won't be within twenty points when the dust clears." Jabbing Lee in the ribs, she teased, "We're going to clean your clock."

"Why should the outcome of the game mean so much to you?" He grinned, protesting, "Anyway, last night you didn't seem too eager to claim the Cardinals. Besides, don't college ladies have more important issues to consider?"

"Ouch! I sense irritation."

"Darn right. Ah, I thought tomorrow's game was about basketball, not a cleaning demonstration."

"Okay, okay. I apologize for a poor choice of words. Would and old fashioned butt whipping sound better?"

"You and your sister are hopeless. What do you think about getting out of here to go get coke or something?"

"Or something? What an interesting idea, don't you think, Sandy? I think Lee's on the right track, but I sure wish I knew what he meant."

My God, she's beautiful. Only the hands of a meticulous sculptor, exercising loving vigilance, could have defined those features.

As he continued to study her, he noted a well-toned physical presence, a contradiction to the totally feminine pose most girls preferred to project. She quietly exhibited a firm, curvaceous figure with just a hint of sensuous excitement. Her outgoing, mesmerizing personality dampened an overt pronouncement of that feature.

Those blue eyes of hers, God, I can't seem to keep from looking into them. Hmm, how would it be to have even, straight teeth like hers?

A further survey revealed a nearly perfect complexion framed by long blonde hair.

Wouldn't it be something to have her for a steady diet? Maybe if I play my cards right....Sure, dream on. Her future's all mapped out.

As the melodic strains again invaded the upper ballroom, he gently took her by the arm.

"Come on, let's dance. I'm tired of arguing."

"Me, too. Um, I didn't support your suggestion because I didn't know exactly where you were proposing we should go."

"Really! You didn't seem to mind when your sister had such a concern about my meaning, though. Maybe with a little less concern about the outcome of tomorrow's game, we could be headed some place where it's a little more private."

"You're absolutely right. I'm sorry I left you hanging out there all by yourself. Forgive me?"

"I suppose."

"Honey, I don't care what we do as long as long as we're together."

"And the mob in here doesn't matter?"

"Not when I'm with you. We could always go in the corner where it's a little less crowded."

"Yeah, and maybe it'll give me the chance to find out who Sandy James really is."

"Later. Right now I just want you to hold me close. Are you going with anyone special?"

"Are you kidding? I've never gone with anyone, ever."

"I think I'll reserve comment until I find out whom you're waiting for."

"I'm not waiting for anybody, at least not until now."

"Hmm, you're going to save yourself for me?"

"Always a step ahead. You really don't expect an answer, do you?"

"No, probably not. It's an intriguing thought, though."

"Fair is fair, Sandy. Now it's your turn. Do you really expect me to believe you're not going with anyone?"

"Sure. I haven't had anyone special in my life since high school, unless you want to count a few dating experiences I've had here."

"Hard to believe."

His eyes fixed on hers, as though seeking a hidden clue.

"What's your real role this week? You must have some reason for being here other than hanging around a bunch of teenagers."

"The university offered me a job to help with the details of hosting this tournament. The money was too good to pass on, so I decided to stay and work. By staying on campus, I was able to offer Jennifer has a place to live during the tournament. I got a bonus out of the deal."

"Meaning?"

"I met you."

"I'm a bonus?"

"Sure. I don't make it a habit to waste my time with just anybody."

"I like spending time with you, too."

"Would you like to kiss me?"

"The thought crossed my mind."

"Well?"

"Isn't it a little crowded in here?"

"Not for me, but, ah, I guess we can always wait until later."

"Yeah. Ah, so, is it only you and Jennifer staying at the sorority?"

"Why do you ask?" She smiled, mildly rebuking him. "My, when you're uncomfortable you cheeks really get red. Okay, you brought it, so look me straight in the eye and level with me. Tell me why you asked."

Glancing at her, suddenly unable to meet her glare, he mumbled, "No reason, I was just wondering."

"Really!"

"Yeah, really. If showing an interest in your activities bothers you, maybe we should drop it."

"Ah darn! Now look what I've done. I've made you angry."

"I'm not angry." Frowning, he snapped, "Let's just drop it. I wasn't trying to be intrusive. I was only trying to ask a simple question."

Suddenly she threw her arms around his neck and kissed him. At first gently on his cheek, she gradually moved on to his responsive lips.

Finally pulling away from the embrace, she whispered, "Is everything all better now?"

She waited patiently for his eyes to finally meet hers.

"It was really nice of you to care enough to ask. By the way, I like the way you kiss. You're passionate without being intrusive. I don't like it when someone jams their tongue down my throat the first time they kiss me."

"I like the way you kiss, too."

A smile started to creep over her face as she pulled him close and kissed him again. When their lips finally parted, she symbolically put her finger to his lips.

"I like you, Lee, ah, I like you a lot. Let's not fight anymore. I'd like to just enjoy the moment." Glancing at the clock on the wall, she muttered,

"Darn, it's already ten-thirty. Won't be long before they close this place down for the evening."

"What about going to your sorority?"

"Not tonight. I don't know about you, but I'm bushed. I have to be at work tomorrow by seven." She smiled, suggesting, "Maybe tomorrow night. How does a small, intimate party sound?"

"Right now, it sounds terrific, but maybe we should wait to decide. You may feel differently after you've slept on it."

"Jen and I'll meet you in the ballroom tomorrow after the game. Um, after we celebrate our victory, maybe we can decide what doing something means."

"Please, don't start the something discussion again."

"Hold me?"

"My pleasure. Ah, I really like you, Sandy."

"Ditto. I'd really like to be alone with you tomorrow night if you feel comfortable with the idea. You're very special, Lee."

Friday night's semifinals brought together all the frenzy and excitement anyone could imagine. In front of nine thousand screaming fans, the North-Lincoln game went down to the wire before the Cardinals prevailed. Forced to accept the unpleasant reality they wouldn't play for the championship, a pall enveloped the Vikings. Later, in the upper ballroom, Lee held Sandy in a close dancing embrace.

"You're very humble tonight. I expected a little gloating."

"Lincoln was very fortunate to win the game. I don't think the cleaners would be a real good place to visit tonight."

"Forget your claim ticket?"

"Uh-huh. It's a good thing the Cardinals won, though."

"Oh? I was sort of hoping it would go the other way, but to be polite, may I ask why?"

"I don't look good in black. Mourning has never been my thing."

"Ah, ha, and it's okay if I do all the grieving, huh?"

"Funerals aren't any fun for anyone. What do you think about getting out of here? I'm tired of sharing you."

"No sense asking Mel and Jennifer how they feel. I think they've been ready to go since they got here."

"Would you like for me to pick up some beer?"
She winked, casually rubbing her hand across the rear of his trousers.
"Course, I don't want to force you into anything."
"It's okay with me. Ah, yeah, sounds pretty good."
"Well, then, it's settled. Go tell them we're leaving. While we're standing here jawing about it, the night's wasting away."

Outside the Erb, Sandy directed, "Jen, you and Mel go on over to the house and get things ready while Lee and I make a beer run." She grinned, casually inquiring, "You drink beer, don't you, Mel?"

"You read my thoughts, but don't take all night. I'm thirsty."

"We'll only gone for about twenty to thirty minutes. Can you wait that long?"

As Lee got into Sandy's car, he teased, "I'm sure Mel and Jen can find something to do while we're gone."

I wonder what Sandy has in mind? Shit, old dear, Lee. Stop dreaming, already! It's one thing to spend the week with her, but it's something else to think she would even consider anything beyond hanging out and sipping a few beers. She's not going to get carried away with her sister and Mel around. Besides, she's already said she has too much to accomplish to get involved.

The car door opening shattered the silence.

"How about giving me a hand, lover. I got a case, think it's enough?"

"Plenty! I'm not a guzzler. A couple will do me just fine."

"Good. A drunk who thinks he's fallen in love really turns me off."

"I wouldn't have to be drunk to fall in love with you, Sandy."

"You stole my line. My house is up ahead."

"Where are you going to park?"

"In the back of my sorority."

"Hmm, your house is the only one showing any sign of life. I'll carry if you catch the doors."

"Before we go in, I have a favor to ask."

"What's that?"

"We may have this part of the campus to all to ourselves, but it would still be a good idea for you to help me keep Jen and Mel under control. I don't relish the prospect of getting arrested for contributing."

"No problem, Mel's like me. After one or two beers, he's done."

"I certainly hope not. I was sort of hoping we could make the evening last awhile."

"I meant Mel and I don't lose control."

Waiting anxiously in the kitchen, Jennifer's impatience boiled over, snagging four bottles of beer before Lee had a chance to set the case of Budweiser on the counter.

"Here, Mel, you do the honors."

"Think Jennifer's acquired a thirst?"

"Like I haven't! I thought you two would never get here with the supplies. Ah, good, Budweiser. It's my favorite."

Lee watched intently as Mel opened bottles.

Hmm, Mel did okay for himself. Jennifer's a knockout, too. Hmm, her body isn't bad either. Other than being a little taller and having darker hair, she could easily pass for Sandy's twin. Um, the biggest difference is the way Sandy handles herself. She's so damned sure of herself. She seems to know exactly where's she's headed and what it takes to get there. I don't know how everything's going to shake out, but...

"Hey, slow down you guys. I haven't even had a sip and you're already ready for another."

"The first one always goes down faster, Sandy."

"Son of a gun, I hope you guys don't expect me to play bartender all night." Meditating, Sandy suggested, "It sure would be nice if we had a cooler."

"The supply run would be shorter. I have one, but I left it home."

"How convenient, Mel. Oh, well, we'll manage. From now on, though, you're all on you own. If you want it, get it for yourself. Is the Oldsmobile parked out back yours?"

"The title has my parent's name on it."

"Ah, ha. You're the caretaker. My parents own the title to my car, too. They've been real good about letting me use it. I've had it down here since my freshman year."

"Don't remind me, Sandy. With the car down here, I've been forced to hoof it. What do you say we take some beer and head on up to the party room?"

"Good idea, Jen. Somehow the kitchen isn't exactly my idea of a great place to spend the evening."

"I'll grab a six-pack unless you think we'll need more."

"Ah, Jen, you might consider slowing down a little. I know Mel wouldn't complain about you getting a little uninhibited, but…"

Sarcastically, Jennifer scornfully chided, "Thanks for reminding me, Sandy."

"Oops, I'm sorry, Jen. That was rude of me. The last thing you need is big sister acting like your chaperone."

"Apology accepted."

"Well, okay, then. If you don't mind, why don't you take Lee and Mel upstairs? I'll be along shortly after I turn out the lights."

In the second-floor dormitory styled room, Lee debated which twin bed to take while Sandy quickly made her way towards the windows.

Hesitating as she looked down on the parking lot, she mumbled, "Still no sign of life in the Alpha Phi sorority. Just in case, I think I'll lower the blinds. Jen, you can set the beer on one of the study desks. What do you guys want, the radio or the hi-fi?"

Lee shrugged, suggesting, "You pick, as long as the music is soft and soothing."

"Good! We'll go with the radio, less maintenance." Fiddling with the tuning button, she groused, "What are you guys waiting for? Find yourselves a place to plant. I'm short of chairs, so grab a bed. Lee, take the one on the left. It's mine. Mel, before you try out the other landing strip, how about dousing the lights?"

A moment later, with only a dim trail of light to follow, she carefully found the way to her bed.

"Not real comfortable, is it?"

"It's not so bad."

"Good, then I'll use you as a cushion."

Grunting, she positioned herself in the cavity of his body.

"Are you comfortable?"

"I'm fine. I like having you close."

"Yeah, but the format of this arrangement is a bit awkward, don't you think?"

"A little. It works when you meet someone for the first time."

"This isn't exactly a first encounter for us, is it?"

"Ah, no."

The projection on the wall across from them featured two figures rapidly advancing towards becoming a unified heap. A prelude to an announcement of intensified involvement, sounds of rustling clothing and unrestrained moans of pleasure, left little the imagination. The heightened excitement had suddenly made the theater overcrowded. Sandy shifted uneasily in Lee's arms, stiffening as the activity intensified.

"This isn't gonna work. I'm tired of pretending not to see what's going on. I'm no goodie-two-shoes, but if they're going to do their thing, I'd just as soon they did it elsewhere. Besides, it's kinda cramping our style."

Suddenly, a hiss escaped the tangled heap.

"Not now!"

"What's the matter?"

"I wanta wait. This isn't the time or the place, especially with them in the room. Damn it, I meant it, Mel, stop it."

"Ah, come on, if we're quiet, they won't even know."

"No, I said. I have no intention of doing it with them in the room."

Abruptly the mass of black light on the wall revealed two distinct forms. A scene depicting rejection featured two shapes, one perhaps making an adjustment to their costume. The forms had suddenly become antagonists.

"Sandy, is there another room we can use? This is too much like parking when you're out on a double date." She grunted, "Shit! Damn it, Mel, you ripped the button off my blouse."

"You're not parked, Jen."

"What's the difference? We still don't have any privacy."

Banging her beer bottle down on the desk, Sandy mumbled, "You're right. It's way too crowded in here. Hurry up and get yourselves decent so I can show to my big's room. Ah, her room will give you plenty of privacy."

"Damn, Jen, this damn thing is stuck. Can you help me?"

Sandy chuckled, suggesting, "Jesus, Jen, give your boyfriend a hand with his zipper. Just make sure you don't get the wrong thing caught in it."

"There, it's okay now. You had your shirt stuck in it."
"Got it! Whew, what about you? Are you done yet?"
"Almost. Damn, I'll have to find a pin somewhere."
"I doubt you need a pin, Jen. You dressed yet?"
"Yeah, but I wish I had a pin."
"It will just get lost."
Sandy opened the door and stepped out into the hall.
"Come on, Jen. Grab Mel and follow me."
Poking her head back in the room, she proposed, "Give me a minute to get the newlyweds situated in the bridal suite. While I'm gone, why don't you crack the window?"

It's gonna be kinda nice to have the room all to ourselves.

A moment later she peeked around the door.
"I'm back. I'll lock this door so we don't have any surprise visitors."
"Where did you put them?"
"Down at the end of the hall. All the way down the hall."

She hesitated momentarily as if debating the next course of action. Apparently content with her decision she turned off the lights, making her way towards the bed.

"I don't know about you, but I've been thinking about what it would be like to be alone with you almost from the first moment we met."

Putting her knee on the bed between his legs, she leaned forward to brace her weight against the wall. Slowly she bent down, lightly bussing his cheek with hers.

"Well?"
"Me, too."

Greeting her lips eagerly, he released some pent-up caged passion.

"Whoa. Not so fast, cowboy. I'm not going anywhere." Studying him coyly, she whispered, "Let's take this nice and easy. No need to rush, we have all night. Strong, silent type, huh?"

"How so?"
"You communicate your thoughts with actions instead of words."
"Sorry. I guess I got carried away."
"This is your first time, isn't it?"
"First time for what?"

"Haven't done it before, have you?"

"Does it matter?"

"Not at all. Ah, I'm not exactly sure what I'm supposed to say or do at this moment. Whatever it is, I certainly want it to be right."

His eyes, having adjusted to the darkness, could finally make out her features. The cosmetic effect of an absence of light amazed him. Radiating mysterious beauty through the darkened accents of her makeup, she appeared almost ghostly. Her image on the wall made the scene appear even less authentic, overpowering and threatening. Maybe she was just a beautiful actress on the stage of a dream. No need for another beer. He was already hallucinating.

"Me either, but I doubt anything could be better than this."

"You're sure easy to please." Adjusting her position, she coyly inquired, "I want you to make love to me."

"Are you serious?"

"Uh-huh. Shhh, I've been thinking a lot about how things might turn out tonight, and what I'd like to see happen."

"And?"

"What I'm feeling isn't just some spur-of-the-moment impulse."

"I don't even know what I'm thinking right now."

"I hope you're starting to think about making love with me."

"But you could have anyone you want. Why me?"

"I don't want just anybody. Is it so hard for you to believe I really like you? You've really thrown me for a loop, in case you didn't know it."

"I have?"

"Yeah. I normally wouldn't even consider doing something like this. It's not worth the risk, especially considering I have everything to lose and nothing to gain."

"You still haven't answered my question."

"I know. Hmm, why you?"

She studied him carefully.

"Maybe it's because of the way I feel and the way I felt when I first met you. I don't really know how else to put it. I really like you, maybe....Look, ah, I know neither of us is ready for anything serious like making a commitment right now. You've got some growing up to do and so have I, but

I'd like to mark a place in the book so someday when we're ready we can pick up the book and move on."

Bending forward towards him, she rubbed her nose softly against his cheek.

"Do you feel it? Can't you feel the electricity when we're together?"

"If you mean the butterflies I'm feeling in my gut, yeah, I feel it."

"Do you believe in love at first sight?"

"I thought that stuff only happened in Hollywood."

"What's wrong with right here?"

"You're not playing games, are you?"

"I don't play games."

She kissed him lightly on the lips.

"How would you feel about getting naked?"

Nervously he exhaled as the two silhouettes on the wall posed motionless. Not unlike two antagonists, one seemed poised to finish off its foe.

"I'd like nothing better, but please, tell me this isn't a hoax. If it's only some prank you have to pull so you can fulfill a bet…I know all the stars aren't exactly in alignment right now, but I'm really hoping there's a future for us."

"They're close, and, for your information, this is far from a joke."

"Would you think I was nuts if I told you I think I love you?"

"Not at all. I feel the same way."

She leaned forward, brushing his lips. The specters on the screen cued and posed, wavered tentatively as they started to merge.

An immediate warm burning sensation invaded his gut, quickly spreading throughout his entire body. Aroused, he moaned when their lips parted.

"I not just interested in tonight. I want more."

"I feel the same, but we have to be patient."

"Meaning?"

"Someday we're going to be an important part of each other's lives."

"Someday. Hmm, could you give me a ballpark figure?"

"In five to six, seven years, tops, I'm gonna come looking for you. If nothing's changed, you're going to have to beat me off with a stick."

"Sounds too good to be true. By the way, how did you know I had never done it before?"

She giggled, nonchalantly shrugging as she leaned back to meet his inquisitive glare.

"Oh, I don't know. You just acted like one."

"Acted like one?"

"A virgin, silly."

"How does a virgin act?"

"Like they've never done it before. Don't worry, it's not a bad thing. In fact, I'm really quite impressed."

"Really. Well, since you're the expert, ah, shit, why did I go and say that? What I said was so unfair. I'm sorry."

The silhouettes on the wall remained motionless, expectant. The more dominant shadow seemed to retreat, shrink a little. Finally it moved, its orb darting back and forth across the screen, a nod of agreement?

"I haven't done it a lot, but I'm not so ashamed of what I've done I would apologize, either. By the way, I think you're very sweet. Innocent, but sweet. Neither of us can change the past. All we can do is work towards controlling the future."

One of the shadows appeared to nod in agreement, the other resuming its confident governing pose. The dark room, momentarily silent, harbored an air of expectancy.

"I guess I'm feeling a little uneasy. I'm, ah, well, like you suspected, this is a first."

"It doesn't matter."

She moved forward, gently brushing her lips across his cheek, stopping her tour at his ear.

"I'd like for us to make a memory tonight. I love you, Lee."

"I love you, too, but I don't want to disappoint you."

"No chance. You have absolutely nothing to prove to me. Let's take it slow and let everything happen naturally."

She gently brushed his cheek again with her lips, the more dominant shadow merging with the bust of the other figure. Slowly shadowy probes engulfed it to form in an indistinct merger.

"Nothing matters other than what we share with each other."

Suddenly two dark rays extending from the more dominant torso formed a T, making indistinct movements on the wall.

"You're modest. It's really nice to be with someone who doesn't have expectations. Um, I can't do this alone." Suddenly rising from the bed, she suggested, "I won't look if it'll make you feel better."

Removing her blouse while studying his fumbling attempts to undo his shirt, a slender tentacle suddenly directed a black indistinct form across the screen, stopped, and then paused as a blob of black light slowly parachuted towards the dark base below.

"Stand up, I'll give you a hand."

"I've got it. Ah, if this is what you call getting naked, I'm not doing a very good job."

He chuckled nervously, casting his shirt aside.

"You're right, it would be easier if I'd stand up."

The shapely phantom again formed a T with its slender appendages. After several indistinguishable movements, the right investigative form of black light moved upward and to the left, paused, and then slowly descended. The left probe repeated the movement again, stalling. Seemingly intent to observe the respondent form finish its task, a chuckle accompanied the blob of light rushing across the screen, parachuting to the base below.

"Hold me?"

He gasped as the figures merged. Exhaling slowly, he reveled in her presence.

"You are so beautiful."

"Oh, my handsome man, you are very flattering. Are you gonna be okay?"

"I'm really nervous. Ah, should I take off my pants?"

Nodding, she began to unfasten her skirt. When the button and zipper had been undone, she slowly wriggled out of it, allowing it to fall harmlessly to the floor.

"Don't be so modest. It's time to lose the shorts."

He nodded, fumbling nervously with the band of his shorts. His fingers responded awkwardly as though they no longer received signals from his central nervous system.

"I like watching you undress."

He blushed, mumbling, "Me, too. I mean, I like watching you undress, too. Ah, I've never done this in front of a girl before."

"You mean you've never played I'll-show-mine-if-you-show-me-yours?"

Chuckling, she encouraged her panties to slide down her legs, settling at her feet.

"There's always a first time."

Undecided about what to do next, he glanced at the heap his fallen clothing had made, nervously fussing with his semi-dormant organ. He gave it a quick tug.

Snickering, she inquired, "Why do guys always do that? Are they trying to make it longer?"

"I don't really know. I was, ah, I was just trying to straighten it out."

Removing one foot from her silky undergarment, she scooped it up with the other.

"Watch, I want to show you something."

Artfully flipping the garment towards the door, in one swift, unhesitating motion, she delivered a perfectly directed pass encircling the doorknob.

"Two. I used to practice making baskets while undressing in my bedroom. Looks like I still have the touch. I'll bet I can still make ten out of ten."

The display on the screen couldn't have portrayed the depth of their emotions or the total significance of the confusing movements manifested on the screen, silhouettes behaving like a piano teacher showing its pupil the proper positioning on the keyboard.

The directed appendage reached out, a lustful voice encouraging, "Touch me?"

He gazed at her inviting mounds, jutting upward like a proud cadet standing at attention in formation. Swollen caps reminding him of ripe raspberries decorated her treasures.

"They're so soft."

"Oh, God, you make me feel so good. You have great hands."

Her gaze suddenly distant, the slow deliberate movement of a tentacle towards the other form's frontal bulge started to make a slow, methodical examination. Like a sculptor working with soft malleable clay, the tentacles molded the offering into an eager surging lance. Suddenly, the slender tentacle darted towards a probe busily tending to a mound of black light, taking it and slowly guiding it downward. A pause, a noisy rush of air, and then a gasp shattered the silence. Another deep breath, a loud rush of air, and then suddenly short panting breaths coming in shallow uneven intervals began to fill the chamber as the confusing activity on the wall indistinguishably revealed the emotion the performers experienced. Imperceptible movements and muffled sounds of pleasure only gave the barest hint of the real plot unfolding.

"I'm almost ready. Oh, Lee, I want you so bad."

He shuddered, leaning forward to kiss her. Greeting her open mouth he responded to the presence of her tongue, making darting tours inside her mouth. Suddenly she pulled away. Taking a couple of deep reassuring breaths, she exhaled slowly, turning towards the bed.

The lead shadow braced its probes on the base of the screen, slowly lowering its configuration to form a complete union. One, two, three upward thrusts, and it finally settled expectantly. Similar to a partially open jack-knife held with the handle and blade facing upward, it was poised and waiting as black appendages reached out to welcome the approaching image.

He tried to steady his breathing pattern, his breaths suddenly irregular and frantic with excitement. Every repressed feeling consumed him, an indescribable erotic wave of emotion overpowering him as he experienced the urgency of the moment. He shuddered again. Engulfed in emotion with temples pounding, a compelling rush of feeling swept over him as soft pulsating folds of sensitive warm flesh gripped him, sending a charged impulse through his entire nervous system.

Merged, the unified silhouettes made no movement as if waiting to be cued for the beginning of the next scene. A whining, creaking invasion of the tomb's serenity suddenly announced the unified duo's performance. Deliberately, conjointly, the forms started making unified movements. High-pitched sounds resembling a poorly bowed violin accompanied the drama as the rendition intensified, shadowy forms culminating their performance as the play neared its climax.

A gasp and then a frantic whimper urged the upper figure on, willing it to move more rapidly. Another gasp and then several unrestrained whimpers of delight pleaded for more. Outside in the parking lot, a motorist gunned the car's engine once, twice and then turned off the ignition. A brief faint report of an engine not completely spent, sounded as a breeze filled the blinds, pushing them away from the open window. Ebbing, the light wind disappeared, sending the window covers crashing against its frame as the waning, whining report of the duel between the bed and frantic pleas subsided.

Silence! The blobs of black light at the base of the screen lie merged in a nondescript configuration. As in death, the unified heap was motionless. Competing sounds having subsided, only an occasional rustle of the blinds and frantic efforts of spent activity to regain normal breathing patterns were detectable. An unexpected interruption shattered the silence.

"I'm sorry, I couldn't last any longer. Sandy, I love you so much."

She sighed, "I love you, too. You make me feel so complete."

"I never expected…."

"You never expected someone to fall in love with you?"

"I, I never expected you to feel the same way I do. Was it as good for you as it was for me?"

"Oh, yes. You were terrific."

She kissed him on the cheek, hugging him tightly.

"You made me feel like a person does when they first start to plunge downward from the top of a roller coaster." Moaning, she continued, "Every wonderful emotion I've ever experienced hit me. I've never felt like this before."

He grinned, probing, "How was the trip to the bottom?"

"Fantastic! I though I was going to wet my pants."

"That would be a real stretch. They're lying over there by the door beside mine. If this is what heaven's like, I want to go there right now."

He kissed her, shivering with an uncontrollable urge to possess her.

"Sandy, when it's right, I want us to be together, forever."

"Me, too."

She exhaled loudly, registering her protest, "Darn, where do you think you're going? Don't leave me now. Ah, shoot! You don't play fair!" She chuckled, innocently inquiring, "What's the matter, is it ready for a nap?"

"I guess it needs a break."

"Is that what it is? Oh, well, I enjoyed your visit. I could really get used to this."

She kissed him, shifting uncomfortably almost hesitantly under the weight of his body.

"Ah would you mind letting me up? I have to go to the bathroom. Um, nature calls."

Struggling to her feet, she rushed to the door. Poking her head out, she started to scope out the hall.

"Oops. I'd better put on my robe. It wouldn't be too swift parading around in the hall naked, would it?"

Covered by the robe, she again opened the door, gingerly stepping into the hallway.

"Good, no intruders! Don't go anywhere. We have some unfinished business to take care of when I get back. Do you want me to bring you another beer?"

"No, I'm good. Two's my limit."

She smiled, studying him intently for a moment.

"I really love you. Someday…."

She bolted out the door.

Wow! Was that good or what? So, this is what it feels like to be in love.

A moment later, she burst into the room.

"I see you decided to crawl under the covers. Good! It's a little cool in here."

Shedding her robe, she joined him in bed.

"You're so warm. How about cuddling me? I'm cold."

"My pleasure. Ah, it was more to me than just doing it. I love you. I really love you."

"Nice to hear, but what you're experiencing now could change. The important issue is, will you still love me tomorrow? That's what really counts."

"Why would I feel differently tomorrow?"

"I don't know. Some guys feel differently as soon as they've done the deed. I think they call it regret. After the conquest, nothing's left."

"I could never feel that way with you. Um, you're not gonna get pregnant, are you?"

She snorted, her soft breasts pressing against him as she started to kiss him on the shoulders and chest.

"Maybe. There's always a risk." She chuckled, teasing, "See what I mean? You're already starting to have doubts. What a break! The first time in the saddle, you become a daddy."

"Are you serious?"

"No, I was only kidding, honey. Strange isn't it?"

"What do you mean?"

"It didn't seem to bother you a few minutes ago."

"Who had time to think? I couldn't seem to think of anything except how good it felt and how much I wanted you." Whispering as he leaned forward to kiss her on the forehead, he pledged, "I still love you."

"I love you, too, and, ah, I couldn't think of anything else, either. You absolutely drove me crazy. I'm pretty sure we don't have anything to worry about, but just remember, everything that happens from now on is out of our control."

"What do you mean?"

She smiled devilishly, explaining, "When you launched your little missiles, the wheels were set in motion. It's not as though we can turn on a vacuum and reverse the process, you know."

Suddenly she pushed herself away from him, staring at him coldly.

"I should start my period tomorrow. Feel better?"

"A little."

"I wish tonight would never end. It's been amazing."

"What do you mean?"

"Well, for one thing, we're so compatible. If I didn't know better, I would think we've had known each other for a long time, a real long time. I feel like I've loved you forever."

Slowly slipping her hand under the covers, she reached for the band of his shorts.

"Wanta do it again? Oh my! Looks like you're already set to go."

He exhaled loudly, probing, "Are we normal?"

"I don't know about you, but I am."

"No, I mean, do you think people fall in love this quickly?"

"We did, that's all that counts."

She put her arm around him, hugging him tightly.

"Kiss me? I want to make love with you all night."

The faint hue of an awakening sun preparing to rise peeked through the haze on the horizon as Lee sat silently in the seat next to Mel.

Hmm, it already five-thirty. The chickens are already up, but the streets are still nearly empty. I sure hope Mel's aunt and uncle are still asleep. Sandy wasn't kidding. We nearly made love to each other all night. Without any sleep, I'm going to be in fine form.

Lee started to squirm uneasily in his seat. Inside, he was starting to wrestle with the ambivalence he felt

Sandy was right. I do feel differently. It's not the same as when I was with her. Part of me wants to scream out, announce to the world what I've done because she's so damned beautiful and wonderful. Making love with her was absolutely out of this world.

He shook his head, staring out the window at the sights passing by, nothing.

I don't know why, but another part of me feels kinda empty. My stomach aches. Who am I kidding? I love her, but somehow....

A chilling fear of apprehension spread quickly through him, enveloping him entirely.

No, I really love her. Everything we did was absolutely fantastic. I just hope she isn't pregnant.

He jerked back to reality as Mel stopped at a red light.

"What?"

"I asked if you have a good time with Sandy tonight?"

Although they practiced an unwritten code of not discussing or bragging about any of their encounters, the question posed by Mel was his way of confirming some of the turmoil he also experienced. Both of their mannerisms clearly indicated Sandy and Jennifer had introduced them to a deeper involvement than ever experienced before.

"Yeah, Sandy's wonderful. How was your evening? Did you have a good time?"

Mel nodded, returning to his silent mode.

Should I, or shouldn't I? What in the hell am I going to do about tonight? Sandy talked about getting together at the sorority again, but I'm not sure. Damn, I sure hope she's not pregnant. I really love her, but

I'm not so sure that I'm ready to get involved like we did last night again. I doubt she is either, especially with all the plans she's made for the future. Maybe we shouldn't have done it. Messing around is a surefire way to get into trouble. Damn, it was so great being with her, but there's no way I could have a relationship with her. Living sixty miles apart with her in college and me still in high school, it wouldn't work. Besides, she seems to be so sure about the future. Hmm, I wonder.

Mel again broke the silence, fuming over missing a light.

"Are you going to see her tomorrow, I mean tonight after the game?"

Lee shrugged. "We're supposed to meet at the Erb. She mentioned going over to her house again."

"That's what Jennifer wants to do. Ah, how do you really feel about getting together with them? Don't get me wrong, I had a great time."

"But?"

"I don't know. I just don't know."

"I feel the same way. A large part of me wants to, but the rest of me, well, I'd almost like to forget about it and pretend it never happened. I think I love her. No, I know I love her. She's really special."

"We're too young."

"I know it. I guess I'll make up my mind when I see tonight."

"Good idea. By the way, how do you think we'll do against Eugene?"

"I really don't know. I worry about Cory Warren. He's something else. I hope we can keep him from scoring."

"You mean, hold him to under twenty points, don't you?"

"Yeah, don't we wish."

"Stopping Cory is the key to the whole game, but I think we might just pull it off."

Whatever concerns they might have harbored disappeared as the exciting battle for third place unfolded, the Vikings rebounding with fury from the disappointing semi-final loss. In the packed old gymnasium, electricity seemed to be in the air. The Viking rally squad must have also caught the spark because they put on the performance of the tournament. Every time-out, every break in the action, the dynamic yell squad sprang into action again spurring their faithful supporting cast to yell even louder, clap with even more enthusiasm.

A great defensive effort controlled Cory Warren's influence on the game, as a strong team effort spurred the Vikings on to earn the third place trophy. At the award's ceremony, after the tournament had ended, the Viking's team captain accepted the third place trophy and, later in the ceremony, made his way to the center of the court to receive acknowledgment for being named to the all-tournament second team. Perhaps more exciting to Lee and the rest of the student body was when the Viking yell queen went out to receive sportsmanship trophy, symbolic of a week of hard work and excellent behavior by the Viking faithful.

Tired supporters gathered at the Erb Memorial for the final dance. While waiting for Sandy, Lee waged a private visceral war.

I'd really like to spend another evening with her, but.... Get real, Lee. Right now, we live in totally different worlds. Maybe the difference in our ages won't mean anything in a few years but right now.... Anyway, I'd kinda like to join my friends at the celebration party.

"I've made up my mind. I've decided to go to the party. I'm going to tell Jennifer when they get here. What you gonna do?"

"I've been thinking about doing the same thing. There's problem, though. I really love her."

"I think Sandy would be pleased to hear that." Jennifer grinned, interrupting, "Here, she asked me to give you this letter."

Surprised by Jennifer's sudden intrusion, Lee felt a sudden infusion of heat encircle his neck.

"Ah, where's Sandy? Isn't she coming?"

Jennifer frowned, disgust framed on her face.

"No, she won't be coming tonight. She told me the letter would explain everything. She wants you read it on the spot." Turning towards Mel, she groused, "Do I look like a postman, or a special delivery runner?"

Excusing himself, Lee quickly headed for the stairs descending to the main floor or the Erb. When he had reached the main floor, he found a soft chair and sat down to privately read the letter. A hint of rejection had already started to settle about him as he began reading.

Wow, three pages. She must put her heart and soul into this.

He looked up, fingering the letter.

She must have, yeah, she got cold feet.

He looked down at the letter again and started to read it.

Thank God she's not pregnant. Whew, what a relief. What's this? Hmm, she's right. Our age difference does bother me. Hmm, she writes, 'You asked, why me? I'll answer by saying it was no accident. You captured my heart. However, the time isn't right. Both of us have far too much to accomplish. Trying to make a relationship work right now would probably ruin any chance we might have for the future. Right now, we live in different worlds. In a few years, if things go right, well, I'm keeping my fingers crossed. All I can ask is for you not to start without me. I promise to return the favor.' Wow! She really loves me. Yeah, meeting up in the future would suit me just fine. I really love her.

When he returned, Mel and Jennifer were engaged in a vigorous debate.

"No, it isn't like that. I would like very much to see you again, but...."

"But, tonight, you want to join your friends at some party."

"Yeah. This entire week has been real special, but it looks like Lee is gonna have to fly solo. I wouldn't feel right about leaving him alone. By the way, you know he really likes Sandy, don't you?"

"The feeling is entirely mutual. I don't exactly know what my sister said in the letter, but if what she told me after you guys left this morning is only half true, well, she's come down with a bad case of love."

She nodded in mild resignation.

"You're right, Mel, it would be better if you went with your friends to celebrate. I just wish Sandy hadn't flaked out because I really wanted to spend another evening with you."

"I appreciate that. You're really special."

"Yeah, I know. That's what everybody tells me. Oh, hi, Lee. Finish reading the letter?"

"All done."

"Mel, you could always come to Portland to see me."

"Count on it. By the way, the game between your Cardinals and Central Catholic was a real classic. Congratulations."

"Thanks, but it was a little too close for comfort. It sure was exciting, though. Lee, so you know, you really impressed my sister. She's a hard one to please, so you must have pushed all the right buttons."

"She's real special. Ah, you might tell her I promise not to start without her. I think she'll know what I mean."

Later, at the boisterous Viking celebration party, Mel looked at Lee through bleary eyes.

"It was too good to be true."

A tear appeared in the corner of Lee's eye.

"As much as I love her, I know the time's not right. We live in different worlds. Someday down the road….Anyway, here's to a great tournament."

With his glass raised in a mock toast, Mel inquired, "Are you okay?"

"Yep, I had the chance to spend a week with a great lady. You know, I even learned a little about myself."

Mel smiled, slapping his friend on the back.

"Who are you kidding? You learned much more. You earned a degree. Anyway, none of that matters any more. We have to decide where we're going to college."

FINGERPRINTS

The approach of graduation signaled the onset of the best, yet the worst of times. Joyful, yet melancholy memories oscillated between a state of cold overwhelming gloom and warm euphoric gaiety. Commencement would soon announce the end, mark the beginning while serving as a connection to things yet to come.

Making a choice between Willamette University and Oregon State College was not easy. The Capital City University was ideal because of its prominent national reputation, and, even with higher tuition, living at home made the cost of a year's education less expensive. A conflicting argument, the Benton County State College had broader course offerings and a huge social advantage. A nostalgic affair harbored since his early teens magnetically beckoned for him to apply at OSC.

I'd do most anything short of robbing a bank to go to OSU. Maybe when I visit the campus Senior Weekend someone will show me a way to earn the extra money it's gonna take to go there. Getting a scholarship is out of the question, but...

The exploits of college men and women were legendary. A sneak preview of the form of behavior rumored to take place most weekends at parties attended by a small group of Lee's senior classmates. Reports of their weekend experiences were filed each Monday.

Ralph Moore occupied a desk next to him in physics. Purportedly the supreme party animal, he was rumored to consume large quantities of beer, wine, or distilled spirits, the variety unimportant. A limiting factor was availability, but he employed supply-and-demand economics to insure the supply was always adequate.

Short in stature, Ralph's thick dark hair framed a permanent impish expression. He had played guard on the football team, was president of a school-sponsored club, Y-Teens, and held an office in National Honor Society. Since he had already committed to attend Oregon State to study pharmacy, Senior Weekend was merely an excuse to engage in another drinking activity.

"I hear you and Jerry Freed are going down to Corvallis for Senior Weekend. How would you like to ride with me?" His eyes sparkled as he continued, "We're in for quite a weekend. College life is a totally different experience."

Grinning, Lee teased, "Not so different for you, huh?"

"My brother tells me the parties at OSC are wild. The booze flows and the women are on the prowl. It's not a matter of whether you'll score, only how often. Those college babes are hot to trot."

"There may be a few like that, but I'm sure they're in the minority. No girl's gonna give it up for the hell of it."

"Seriously, most women at that cow college are horny as hell."

"If only half of what you say is true, it's still a big bite for me to swallow. How's the expression go?"

"I know, seeing is believing. Anyway, my brother invited me to a party at his fraternity after the dance OSC is sponsoring for us."

"What house is he in?"

"Phi Delt"

He winked, a sly smile etched on his face.

"I'm sure Greg wouldn't mind if I brought guests. He'll even set us up. Ah, you know what that means, don't ya?"

Smirking, Lee challenged, "Enlighten me."

"We'll probably get a little."

"Very little. If scoring is a sure thing, why have dates? Won't all those horny women be milling around like eager cows ready for the prod?"

"You're such a skeptic, Lee. Well, are you gonna ride down with me or not?"

"I guess we've nothing to lose. Ah, I wouldn't plan on selling your story to the locker room press. Most of the guys who brag about getting laid are the same ones who sneak a peak at the *Playboy* centerfold while they're waiting their turn at the barbershop."

"Go ahead and make fun. All I know is what I've been told. My brother, Greg, does quite well for himself."

"And I'll bet he's willing to share, huh?"

The appropriation of contraceptive paraphernalia required nerve and boldness. The anticipation of purchasing them was terrifying, but approaching the pharmacist to procure them was worse.

Ralph wheeled his 1950 Studebaker, a maroon two-door sedan, into a parking space located near the front of the Capitol Drug Store. Located across from the super market where Lee worked part-time, it was an all-purpose drug store located in the Capitol Shopping Center. Ralph's car had barely come to a stop before a heated argument over who would buy began.

"One of you guys go in and buy them. I need to stay with the car."

"I think Lee should get them because he works in the shopping center. He knows his way around better than we do."

"That makes no sense, Jerry. What has working at Borg's got to do with who gets to buy the damn things?"

Jerry's face paled.

"If my dad found out I bought a rubber, I'd never be able to set foot in my house again. You're the man for the job."

"He has a point. You're the man, Lee."

Frowning, Lee demanded, "Tell me again why you think we're going to need them. While you're at it, tell me why you have to stay with the car. Are you afraid it will run away? Personally, I think your chicken, Ralph."

Ralph snorted contemptuously.

"You must be kidding. How do you expect to score without protection? Chicks won't let you make it with them if you don't produce a rubber."

"Why?"

"They don't want to get knocked up."

"Okay, I'll buy the damned things, but you guys owe me big time!"

"Make sure you get Sheiks. They're the best."

"How do you know? Are you basing it on experience?"

"Go on and get in there and buy the damn things so we can hit the road for Corvallis. Just don't buy French Ticklers. I don't want the women following me home after I get through with them."

Lee shrugged as commenced a less-than-determined stroll towards the drug store.

What in the hell am I doing listening to him anyway? I'll bet he wouldn't know what to do with a rubber if the opportunity knocked him in the head. If a girl did agree to do it with him she'll probably have to put it on for him.

Fidgeting with the loose change in his pocket, he entered the store still pondering how he would approach the task. Suddenly his stomach leaped into his mouth, panic setting in.

Damn, just my luck! The clerk's a woman! No way I can ask that sweet lady for rubbers. What in the hell am I going to do? If I go back to the car without them, Ralph will never let me forget it.

Unable to find reasonable justification for not fulfilling his mission, he approached the counter where the middle-aged lady was busily working on an order.

The clerk who could easily have passed for his mother smiled pleasantly.

"Can I help you?"

"I'd like to speak with the pharmacist, please."

"Certainly, I'll get him for you."

Middle-age had encouraged a hint of fullness to capture the clerk's still shapely body, her short brunette hair peppered with faint hints of gray.

Who am I kidding? She knows why I came in here.

In the back, the druggist, also sporting faint twinges of gray hair, labored feverishly to fill an order for a prescription. He appeared to be in his late forties or early fifties. Possessing strong but not dominating features, he could have easily passed for Lee's dad except for a few extra pounds he carried. Operating in the enclosure behind the counter, he nodded knowingly when his assistant informed him of the request. As he looked up and peered over the top of his spectacles, the softness quickly disappeared from his face. Carefully surveying his customer, a thin smile slowly appeared on his face as he hurried around the end of the enclosure to reach the counter.

"What can I do for you, son?"

"I want some rubbers!"

Jesus, Lee, why not just announce it to the entire world?

The amused drug master feigned a serious fatherly posture, understanding, dark brown eyes twinkling.

"Hmm, what brand would you like?"

"What brands do you have?"

Maybe I should have just asked for Sheiks and been done with it.

His neck started to get warm as a strained, flushed expression froze on his face. Waiting for the pharmacist's response, time seemed to stop. He could feel all eyes fixed on him, boring into his flesh.

Does his nametag really say GODFELLOW? What have I gotten myself into?

Lee scarcely heard the elderly man's patient recitation of the numerous options, struggling to overcome his extreme discomfort. The environment of the drug store suddenly resembled the sweltering chambers of a food processing plant during the hottest portion of the summer.

"What?"

"I said, Sheiks are very popular."

"I'll take them, the Sheiks."

The kindly gentleman nodded. Apparently not eager to end the banter with his young customer, he released the drag to encourage the battling prey to make another run.

"Have you ever considered French Ticklers? They provide maximum stimulation for your mate."

"No, I'll take the Sheiks."

"As you wish. How many?"

"I don't know, ah, how do they come?"

"Um, they come by the unit or in tins of three. It's more economical to purchase them by the dozen unless you have an abnormally heavy schedule planned. In that case, I would recommend buying a gross."

From the other end of the counter, a muffled snort followed by a stifled snicker punctured the hushed environment. The pharmacist's clerk quickly turned away, escaping to the back room. Mr. Godfellow smiled sympathetically.

"Perhaps you should consider a tin. It should be enough to get you by. Um, that'll be sixty cents."

Lee nodded, quickly producing the exact change.

Jamming the purchase into his pocket, he mumbled, "Thanks, I won't need a sack."

Turning, he scurried towards the door.

Thank God that's over, what a fuckin' nightmare.

By the time he had reached Ralph's car, his pace slowing, his confidence had returned. He heaved a huge sigh.

Now it's my turn to have a little fun. Ralph will want to know every detail. Hmm, the longer I make him wait, the more impatient he'll become. Okay, you asked for it, Ralph. I'm gonna really enjoy watching you squirm.

"Did you get them? What brand did you get?"

With a disinterested movement of his shoulders, Lee eased into the back seat. Slowly withdrawing the evidence from his pocket, careful to conceal the brand, he flashed them rapidly in front of Ralph's inquisitive glare.

"Oh, you mean these? You didn't think I would come back empty-handed, did you?"

Playfully he jammed them back into his pocket.

"Did you get Sheiks?"

"Funny you should ask. Um, after an in-depth discussion with the pharmacist and his assistant, and realizing obtaining the maximum sensual gratification for both partners is essential, I decided to try a brand guaranteed to produce ultimate satisfaction. Apparently, the little number I selected delights women, absolutely drives them crazy. Ah, the clerk, she told me it heightens their sexual pleasure and increases their enjoyment."

"The assistant was a woman?"

"Not bad looking, either."

"You're shittin' me. You discussed rubbers with a woman?"

"Who better to know? She said next to her husband going bareback, she liked the brand I selected the best."

"No way! She didn't say going bareback, did she?"

He again withdrew the tin of Sheiks. Fingering the container, he continued to conceal the brand.

"Yeah, she encouraged me to buy a dozen or more of these little rascals, but I decided we could get by with a tin. Besides, I didn't have enough money for a dozen."

"A tin? How much did it cost?"

"Oh, the usual."

He winked at Jerry and then turned to face Ralph's inquisitive glare.

"Let's get a move-on, Ralph, or we'll be late for check-in at OSC." He yawned, and then covering his mouth like a trombonist uses a mute while, he mumbled, "The clerk gave me the impression she has tried them all."

"You're kidding me. I'll bet she keeps her old man happy."

"Or worn out. I wouldn't mind if she gave me a few lessons."

Thrusting his arm towards the back seat, Ralph snapped his fingers, demanding, "Give 'em to me! I wanta see if you got the right brand."

"What's the right brand? I sure hope it's French Ticklers. The clerk was adamant she would rather do without sex than fuck with a glove."

"She didn't really say fuck, did she?"

"After listening to her list of reasons for using them, she convinced me the Ticklers are absolutely the best. Ever see one?"

"Ah, sure, lots of times. They look like all the rest."

"All except for the little strand of rubber attached to the end. It's used to stimulate your partner."

"How do you know?"

"She showed me. By the way, what size did you want?"

"They come in sizes?"

"Sure, some guys have bigger equipment than others. If you want protection while still enjoying the experience, you have to get the right size. Ah, what the hell. I guess I might just as well let you see 'em. I sure hate to disappoint you, but I couldn't afford the Ticklers."

The amusing scene inspired Jerry to join the fray.

"It's not too late, Ralph. You could still go in and get some. If those gals at OSC find out we brought French Ticklers, they'll probably line up in groves."

"Yeah, the lady did suggest we try 'em our next time out. She said something about making it easier for a woman to climax."

Impatiently Ralph snatched the tin, eyeing it carefully as he opened it.

"Hmm, Sheiks. You only got three?"

"What'd you expect?"

"Damn, now I wish you'd bought the Ticklers."

Lee shrugged, proposing, "It's not too late. Go on and get 'em; we'll wait. Ah, or you could always wait and pick some up when we get to Corvallis. Nobody knows who you are, there."

"I'm going to put mine in my wallet. You guys better do the same. You don't want anyone to catch you with one of these, trust me."

Lee chuckled, inquiring, "What're they gonna do, arrest us for possession?"

Ralph started the car and angrily jammed it into gear. Squealing out of the parking lot, he was soon on the street leading to the Marion Street Bridge.

"I'm taking the back way, through Monmouth, if it's okay. Maybe you have an opinion about the route that I should take, too. "

"No, you're the captain. Ah, I'm gonna disappear for awhile to dreamland. I didn't get much sleep last night." He chuckled, continuing, "I was thinking about all the hot women I was gonna meet this weekend."

He placed his hands behind his head and leaned back.

Let's see. We've got about an hour before we get to Corvallis.

Patting his pant's pocket, he smiled smugly.

I won't need this. The party's probably gonna be a bust, anyway. None of the hot mammas Ralph talks about is gonna give up just for the hell of it.

Closing his eyes, a smile spread slowly across his face.

Nothing will ever match the week I spent with Sandy. I may never see her again, but.... Nah, whom am I kidding? When she graduates in June, she's gonna travel for awhile and then go to law school. I still have four years left. What the hell, maybe down the road, somewhere, we'll meet up again. What was it she said? Oh, yeah, I'm not supposed to start without her.

He folded his arms across his chest.

Someone like Sandy only comes along once in a lifetime. I sure wish I could see her again and tell her how much I love her. Damn, I still have to get a date to the prom when this weekend is over. I wonder if the gal from South Salem Mel's fixing me up with would be a possibility?

He slouched lower into the seat as sleep started to overtake him.

Mt. Broken Top was indescribable. Nestled in the beautiful Cascade Mountain Range, the beautiful adjunct rested conformably near the Three Sisters Mountains. In full view of Mt. Jefferson and Three-Fingered Jack, it showcased a panoramic splendor no painting or photograph could adequately depict. God seemed to have moved his brush across the horizon, experimenting with different color combinations, different landscape designs. A first hunting trip, unsuccessful, was educational. Besides reviewing a science lesson on the effect of atmospheric pressure, he was introduced to some firsts. Spam, three sandwiches gave testimony to how good something tastes when you're hungry. Water rushing down a mountain in a crudely formed stream, cold and sweet, was ice cold and the best he'd tasted. Endless distances are covered when stalking wild game. When a destination was reached, another always beckoned.

Lee jerked to attention as Ralph slowed the old Studebaker to Monmouth's posted speed limit.

"Don't anyone blink or you'll miss this place."

"I thought you were asleep."

"I was. Ah, hardly seems big enough to have a college located here."

"Nope, there's not much to Monmouth. Um, Corvallis is only fifteen minutes away."

"Good, ah, don't you think we should get our signals straight?"

"You're probably right. We won't see much of each other until tomorrow night. You guys still want to go to the party?"

"Sure, why not. Ah, I still find it tough to believe we're all going to get set up with gals who are hot to trot. Buying these damn rubbers was a total waste, don't ya think, Lee?"

Ralph protested, "One may not be enough. Yep, we're in for one hell of a weekend, but if you don't want yours, I'll find a use for it."

"What are you gonna do, blow it up like a balloon? Why don't you check out the party again? I'm not real excited about going on a wild goose chase."

"Okay, okay, Lee. When I know all the details, I'll let you know."

"I thought the party was all set."

"It is. I'm just gonna confirm the time and stuff."

He turned onto Ninth Street.

"You guys better get out the letter giving directions to your living quarters."

"I have mine right here. Do you want me to read it to you?"

"Yeah, sure. Oh, Jerry, I'll drop Lee off first and then you can guide me to where you're staying. Okay, Lee, anytime you're ready."

"Take a right turn at Harrison and go two blocks. Then, you'll want to take a right. Ah, it's the second house on the right."

He studied the stately two-story houses framed by tall majestic leafy maple and elm trees lining the street.

"My God, this looks like a scene you would expect to see on a postcard. This place is flat out beautiful. Ah, there's Harrison. Take a right."

"Is Corvallis always this quiet?"

Ralph chuckled as he pulled in front of the Kappa Alpha fraternity.

"Nope. Like any college town, It has its highs and lows. Just wait until tonight, Jerry. This place will be hopping. Well, Lee, we'll see you tomorrow afternoon. I'll fill you in soon as I find out something."

Lance Deal did nothing to tarnish the image of a fraternity man. Cool and ruggedly handsome, Lee's big brother for the weekend didn't lack for social assurance or presence.

Lance extended his hand to greet his guest.

"I've been expecting you, Lee. Welcome to the Kappa Alpha house. Come on, let's get your stuff, and get you bunked down on the sleeping porch." He stopped to light his pipe and then continued, "The sooner we get you settled, the sooner we can hit the basement where we're holding the festivities. I don't know about you, but I'm long overdue for a brewsky."

"You always start this early?"

"Shit, man, most of the guys start on Thursday. I don't turn down the opportunity to party, but I can't get started until Friday afternoon. Yep, engineering's a bitch. I also like to reserve a little time for my woman. Ah, there's an empty bunk."

Again re-lighting his pipe, he paused to survey the sleeping porch.

Grinning, he continued, "It's gonna be a great weekend, ah, it could turn out to be a one-timer. Hope you're game."

"I'm not much of a drinker, but don't let me stop you. I'll just tag along and meet the guys in your house."

In the basement, Lance introduced Lee to Trent Roque.

"I have Senior Weekend duties to attend to, so I'm gonna turn you over to Trent. Trent's a newly initiated member."

Trent nodded proudly.

"Yeah, I survived last year's visitation and even managed to get initiated. Ah, Lance's my big brother."

"Anything Trent tells you is gospel. I'll be back."

"Lance is one of the coolest guys in the house. He's always in center of things. Ah, you should see his woman."

"I've noticed. I mean, yeah, Lance is pretty cool."

"Hmm, want a beer?"

"Not yet. I have to work up a thirst."

"I don't drink much, either. Why don't we grab a spot in the corner over there and just sit back and enjoy the music?"

"Sounds good."

Lance is probably one of the leaders of the house. Trent's right, Lance is cool. I think he knows it, though. Suppose he enjoys smoking his pipe or is it just a status symbol? Hmm, so his woman is a real looker. Why doesn't that surprise me? I can't wait to see her.

He continued to take in the activity surrounding him.

Hmm, I wonder what Lance meant when he said this weekend is going to be a one-timer?

Trent interrupted, "Here comes Lance. He must be through getting everyone settled in. Lee, I really enjoyed hanging out with you, but I've got to hit the books. I have a big exam in chemistry on Monday."

Lance nodded as Trent left.

"Studious type. Um, well, did he fill you in on college life and answer all your questions?"

"We didn't talk much. I've just been enjoying the music."

"Uh-huh. On the weekends, we like to let down our hair and zone out with a few Bs. Tonight's just a snapshot of what it's usually like."

"Hmm, it sounds kinda cool. So, when do you see your woman?"

"It has its moments. Lea and I usually go out on Wednesday and Saturdays. Sure I can't get you a beer?"

"You know? I think I will join you. I've finally worked up a thirst. Besides, I can't let you finish off the keg all by yourself."

"It's about time. Shit, man, you're about two hours behind."

"I'm not in a race. Anyway, I doubt I can match your capacity."

"Uh-huh. Beer drinking is a lot like training for a sport. Yep, it's all about getting in shape and getting in lots of practice. Ah, I'll be right back. I'll fetch the brew."

When he returned, Lance handed Lee a cup of beer.

"Have you seen the schedule for the weekend?"

"It's in my bag upstairs. Will I need it?"

"Nah, I have it memorized. You just leave everything up to me. I'll make sure you hit all the important events and see all the essential stuff. You know about the orientation dance tomorrow night, don't you?"

"I knew there was a dance."

"Maybe I'd better tell ya about the schedule before I start slurring my speech. I'm already starting to feel no pain. Whew, when we get up in the morning, I'll take you over to the student union for the orientation. They're going to give you a tour of the campus and explain all the academic stuff the administration feels you need to know. After the orientation is over, we're going over to meet your date for the dance."

"You've already set me up with a date?"

"Certainly! My woman takes real good care of me."

"Do you know anything about her, I mean my date?"

"Nope. Ah, I wouldn't worry. Lea's an Alpha Phi. They have some of the best-looking women on campus." He smiled, revealing, "Besides being absolutely beautiful, she's very discriminating. If Lea has anything to say about it, you'll do fine. Did I tell you we're gonna double to the dance?"

"You seem to have everything organized."

"Only the best for you, my man. Hey, I'd better hit the keg for another brew before one of us dies of thirst."

"I'll go with you. Ah, where they holding the dance?"

"Gill Coliseum. Ah, thanks for not asking if your date goes down. Lea wouldn't share, even if she knew. She's like most women. They get a little edgy if anyone suspects their giving it up. Hmm, suppose they have a code prohibiting them from sharing such information?"

"Could be. Maybe they don't like to advertise."

"Oh, yeah, I'll take you by the sorority tomorrow after the campus visitation and introduce you to your date, or did I say that already?"

"You covered the schedule quite well."

The next day, just before noon, Lee ran into Jerry in front of the student center.

"You enjoying the campus tour?"

"It's okay. Everything's all set for tonight."

"You must have seen Ralph. So, he wasn't full of bullshit after all."

"Everything's looking pretty good."

"I'll defer judgment until later."

"He told me to tell you we're picking you up after the dance at the place where you're staying." Grinning broadly, he excitedly announced, "I scored big!"

"How so?"

"I got set up with a real beauty from Portland, ah, the dance tonight?"

"Oh, yeah. Um, how do you know?"

"I met her about an hour ago. She's flat out gorgeous. Um, I guess Ralph didn't do too badly, either. Seen yours yet?"

"Not yet. After I check out the financing possibilities, my big bother's going to take me by the Alpha Phi house to meet her."

"Oh, yeah. Don't forget to bring your rubber. Ralph's brother says all the gals drink like fishes and fuck like minks down here."

"Hmm, interesting combination. Well, okay then. I'll try to contain myself until tonight."

Lance met up with Lee at the admissions office.

"Let's go big guy. We've got an appointment to meet your date." He smiled. I love Lea to death, but sometimes she drives me crazy. If everything isn't planned to the last little detail, she won't be happy. Ah, you know what that means."

Lea enhanced the image most sorority girls have for being good-looking and socially adept. Although not in Sandy's class, she was still crème de la crème. Greeting them amid a chaotic environment of girls buzzing around like bees in a hive, she shook her head in disgust.

"Will you look at this scene. All these little bitches are either completely consumed with getting ready for the dance or trying to soothe their fuckin' nerves because of a screw-up with the dating arrangements. Excuse my language, but I hope you aren't as fucking particular as they are. Lee, it is Lee, isn't it?"

Lee nodded, his face flushed.

"Please don't say anything when you meet Sally. Wink if I did okay. By the way, Sally's from Medford."

Sally was cute and had a body worthy of a second glance. Within a moment of winking, he wondered if his eye hadn't flinched a bit too soon.

"Are you planning to attend OSC next year, Sally?"

She shook her head, intent to maintain the relationship she had formed with the carpet in the spacious Alpha Phi living room.

"Is that a yes or a no?"

"My parents want me to go to the University of Oregon."

"Where do you want to go?"

"I don't know."

"Hmm, it's a tough decision. I'm debating between OSC and Willamette. Ah, you can't go wrong by going to the U of O. I have a friend who's going down there. Um, it has a beautiful campus. I was down there for the state basketball tournament."

She nodded.

"Yeah, this friend of mine and I stayed in Eugene during the tournament. Ah, we stayed with his aunt and uncle. They were really great about allowing us to pretty much come and go, as we wanted. Yeah, I met some neat people and saw lots of basketball. What about you?"

"We drove up for the games when Medford played."

"You didn't catch any of the after-game stuff, then."

She shook her head.

The damned carpet is tough competition. Holy crap, she must be planning to enter the quality control field. I wonder if she's actually trying to find out the number of fibers per square inch or if she has a physical deformity.

"Well anyway, the dance should be fun."

Lance interrupted, "I think we'd better go. I don't know about you, Lea, but I've got tons of stuff to do. You know what I mean?"

"Good idea. Um, I'll see you later, Sally. It was nice meeting you."

On the drive back to the fraternity, Lee finally broke the silence. "Sally's a little shy."

"Really! I was thinking more along the lines of socially inept. Oh, well, you didn't have plans to fall in love, did you?"

At dinner, the chairman, George Meek, rang the dinner bell to get the attention of those assembled in the dining room.

"Ahem, the traditional processional will be held in the living room. As usual, it will precede the crowning ceremony."

Muffled snickers erupted in the background.

"Hold it down, guys. We don't want to give our guests the wrong idea. Anyway, everyone intending to enter their date in the pageant needs to register them with Lance. Don't forget you have to escort them around the center of the living room so we can all get a good look. As always, we want to select the best candidate. Our reputation is on the line, so take the voting seriously."

Lee nudged Lance, inquiring, "What's everyone laughing about?"

"Just a minute, I'll explain."

"Do you have something to add, Lance?"

"Ah, George, aren't you forgetting something?"

"Oh, yeah, thanks, Lance. Our visitors can't enter the contest. This event is sort of like a Miss College America contest."

Someone from the back of the room interrupted, "That's not fair. What if one of our guests has a qualifying date?"

Lee again nudged Lance, probing, "How do you determine who's eligible?"

"Your date wouldn't stand a chance."

Lance again rose to his feet.

"Ahem, Mr. Chairman, if I may?"

"The floor's all your. Maybe you can offer a better explanation."

"How's it going to look on campus when the word gets around we have to invite outsiders in for this contest? In the past, it's been open to members only. Um, George, since this contest is new to our guests, why don't you tell them what they're supposed to do?"

Someone from the back shouted, "Shit, Lance, all they have to do is sit back and enjoy the parade. This event's like a 4H show. All of the stock is USDA prime."

The assembled membership exploded in laughter.

Again nudging Lance, Lee questioned, "Do I want to know what's going on?"

"No, this is one time when you can have too much information."

At the end of the evening, Lance and Lea took Lee back to the fraternity after he'd said goodnight to Sally.

"Thanks for dropping me off, Lance."

"You have any plans, Lee? Lea and I were hoping to grab a few minutes alone."

"I'll be fine. I have plans to go to a party with some friends."

"You be careful, Lee. Don't do anything Lance and I wouldn't."

"Good, Lea, now he'll get thrown into jail for sure."

"It was a pleasure meeting you, Lea. Thanks for everything."

"I'm sorry about tonight, Lee. Sally wasn't much fun, was she?"

"She was okay, just a little shy. Ah, thanks again."

"Well, anyway, I guess we found out looks aren't everything."

"Yeah. We found out at the crowning ceremony, didn't we?"

In the graveled parking lot of the Phi Delt house, Ralph approached the gatekeeper.

"I'm Greg's brother. He invited us to the party."

"Craig? Sure you have the right frat? This is the Phi Delt house."

"No, Greg Moore. My brother's name is Greg Moore."

"Oh, so you're Greg's little brother." The burly gatekeeper turned to his partner, announcing, "Kurt, this is Greg's brother. Did Greg say anything to you about inviting anyone to the party?"

"Nope, not a word."

"Are you sure? I talked with him this morning."

"Never said a word. I should know, I'm his roommate."

"Come on, guys, you know Greg. He probably forgot."

The mammoth gatekeepers closed the circle, glaring at the intruders.

"Let me spell it out for you, fella. Greg didn't say anything to us, and, even if he did, we don't open these parties to the entire campus. Sorry for the mix-up, but it looks like you're out of luck."

"Would it do any good to give my brother a call?"

Kurt turned to his companion.

"Ah, maybe you can get the message across to these guys, Jim. I'm not having much luck."

"Okay, okay. There's no need to get hostile, we're leaving."

In Ralph's car, Lee smugly inquired, "Well, Ralph, what now?"

Pealing out of the parking lot, Ralph grumbled disgustedly, "Looks like Greg screwed us. What the hell, let's head for Albany. I know a place where there's lots of action. What do you say?"

"Well, we've come this far. What time is it anyway?"

Jerry interrupted, "A little after eleven. Isn't it a little late?"

"Late? Things are probably just getting warmed up. It'll only take us about ten to fifteen minutes to check it out."

"Do you know where you're going?"

"Yep. Let's see. If I remember, I turn right at the end of the next block."

He slowed his car, looking around.

"I'm not used to coming this way. Maybe I'm supposed to turn left. Ah screw it. I'll take a right."

"Fuck, Ralph, you turned the wrong way. We're heading into a parking lot. Are you sure you know where you're going?"

"There's an exit. I'll go around the block and head south down the street I should have turned onto."

"What then?"

"Well, then I'll take a left and head towards the Willamette River."

"I hope you don't plan on swimming to Albany."

Jerry shook his head, mildly protesting, "I'm thinking the action in Albany probably won't be any different than the stupid frat party. If you don't find the way out of Corvallis pretty soon, I think we should call it a night."

"Ah ha. There's the street I'm looking for. If I take a left, we'll be on the street leading to the street leading to the bridge crossing the Willamette River. Ah, shit! I made the wrong turn."

"Ralph, you took a one-way street. You're going the wrong way."

"No shit! Tell me something I don't already know, Jerry."

Police lights and a shrill siren made a confirming statement.

"Quick, check and see if the box on the floor is covered with the blanket."

"What in the hell's in the box, Ralph?"

"Don't worry about the box. Act like you don't know anything about it. Is it covered?"

"Yeah, but what's in the box?"

Lee turned towards the back seat to take a better look.

"Damn it, Lee, turn around and face the front. Let's worry about the fuckin' box after we're done with the cops. Ah, make sure you pocket the rubber in your wallets. We sure as hell don't want the cops to find them."

The officer approaching Ralph's side of the car flashed his light into the front seat and then slowly started to survey the back seat. Suddenly his light stopped. Nodding to a second officer approaching on Lee's side, he tapped on Ralph's window.

"Roll down your window and get your driver's license out. Driving a little erratic, weren't you?"

"Why, was there a problem?"

"I've been following you for the past few blocks. You were either completely turned around or you've had one too many, son."

The second officer, now standing on Lee's side of the car, started flashing his light around the interior of the Studebaker.

"Wha-cha-all-got under the blanket on the floor?"

He tapped on the top of the car with his left hand as he continued to flash the light around the inside of the car.

"Hey, you in the back, get out of the car so I can take a closer look."

"Sure, what, what ever you, say, say, Officer."

"Go stand in the front of the car so my partner can keep an eye on you. Um, what do ya suppose I'm going to find under the blanket?"

"Beats me. I, I didn't, didn't even know it, it was there."

The officer braced himself on the seat as he reached for the blanket.

"Hmm, Blitz. Well, will you look at this, Roy! It looks like these lads were headed for a party."

Frowning, Roy turned towards Ralph and probed, "Well, Ralph, what about it?"

"Okay if I get out of the car to take a look?"

"I suppose you're going to tell me you haven't seen the case of beer in your back seat before."

Ralph mumbled, "Those bastards! The Delts must have planted it in my car." Turning towards Officer Roy, he continued, "Believe me, I don't have a clue how the beer got in my car. I don't even drink!"

"If you don't drink what in the hell were you doing at the Delt house tonight? They don't host Coke parties, ya know. What ya think, Carl? Should I cuff him?"

Roy slowly reached for his handcuffs.

"Carl, you cuff the other two. Looks like we'll have to take these lads downtown for a little visit to the country club."

Carl nodded, announcing, "You're under arrest for possession, son."

He turned towards Lee, reaching for his second pair of cuffs.

"Yeah, you two lads are under arrest, minors in possession of alcohol. Shit, I left my second pair of cuffs at the station. Hey, Roy, do ya have another set of cuffs? If not I'll cuff these two together."

"Here, catch." Roy turned to Ralph, and gruffly inquired, "Your keys in the car?"

"Yeah, they're still in the ignition. Why do you want to know?"

"So I can lock up your car."

"What happens then?"

"We take you in for processing. Afterwards, if you can't make bail, I guess we'll have to impound it. Isn't that about it Carl?"

Ten minutes later at the police station, the officers paraded their prisoners before the desk sergeant.

"Got three more felons for ya, Sarge."

"Caught some more OSC students under the influence, huh?"

"No, high school kids down for Senior Weekend. Roy and I found a case of beer on them. It was in their car under a blanket."

"Hmm, off to a good start are they?"

"Got that straight, Sarge."

Carl eyed his prisoners coldly.

"Officer James is gonna fingerprint ya and take your mug shot. While he's getting ready for you, I'll take you into the room over there to wait."

The adjacent room contained a table and four chairs. Stark!

"You Lads take a seat. One of us will be back in a few minutes to take one of you in the other room to be booked. It shouldn't take too long get your mugs shot and make a copy of your tracks."

"Tracks?"

Ralph shook his head in disbelief, snapping, "Fingerprints, dummy."

Roy smiled, confirming Ralph's insight, "Yep, one for each hand."

At last fingerprinted, Roy looked at his partner and then at the desk sergeant.

"They're entitled to a phone call. Should I take them to a cell or let them make the call first?"

"Can you lads make bail?"

Ralph frowned, looking around the room for some indication from either Lee or Jerry.

"Depends. How much?"

"A hundred each. You're each entitled to make one call if you don't have the money. If I were you, I'd make it a good one. Otherwise, you'll end up spending the night in our country club."

Jerry mumbled, "I haven't a clue where I can get that kind of money."

"Same here, what about you, Lee?"

"Let me see. First, there was the favor I did for you in Salem, and now I'm expected to bail us out of jail. You sure expect a lot from a dummy, isn't that what you called me, Ralph?"

Roy interrupted, "You lads could always call your parents. I'm sure they don't want you to spend the night in jail."

Chuckling, Carl appealed, "Ah, let 'em be, Roy. Some parents don't take too kindly to late night calls."

Lee got up and headed for the phone, announcing, "I'll call Lance."

A few seconds later, Lance answered.

"Lance, this is Lee. Ah, I'm in jail. Suppose you could come down and bail me out? Ah, me and two friends?"

"How much?"

"Three hundred dollars."

"It's gonna cost ya."

"Could you hurry? I don't want to spend any more time here than I have to."

"I'll be right down. Ah, I want all the details. They guys in the house will want to hear all about it, too. Ah, you can give us a report at breakfast."

At noon the next day, after stowing his gear in Ralph's car, Lee responded to Lance's extended hand.

"Thanks for everything, Lance. You didn't lie when you told me this weekend would be a one-timer."

"The first time I saw you, I thought you were a just another high school kid not knowing up from down. I was wrong. You are a real piece of work. I meant it when I told you we'd have a place in this fraternity for you. Oh, ah, we'll have a beer together when you come down for court."

Just then, two girls walking in front of the fraternity were assaulted by two of Lance's fraternity brothers. Spraying them with cold water from a hose, they hooted with delight as the girls retreated down the sidewalk until they were out of range of the powerful spray.

Lee shook his head in amazement.

"Your house has a passion for making impressions with the ladies, Lance."

"Yeah, and they love every minute of it."

On the outskirts of Salem, Ralph shattered the silence.

"So what did you guys think of Oregon State?"

"I'm still leaning towards Pepperdine."

"There's no way I can afford to go there. I guess I'll have to face reality and go to Willamette. Ah, I don't know what your living organization was like, but mine left a little to be desired. Are all the fraternities like this at OSC?"

Ralph grinned, teasing, "What's the matter, Lee? Don't you like having a little innocent fun?"

"You call spraying women with a hose or having a ugliest woman contest having fun? Don't get me wrong, Ralph, I would like nothing better than to come to college here."

"But?"

"I just can't afford it. If I could, I doubt I'd join a fraternity."

"Don't be too quick to judge. Not all fraternities are like the one you saw this weekend."

"Well, we'll just have to see. Anyway, I don't know about you, but I still have to rustle up a date for the Junior-Senior Prom."

A WHITE SPORT COAT AND A PINK CARNATION

The importance of the Junior-Senior Prom, significant, motivated many to initiate agonizing efforts to find the perfect date. Songs, movies, and more chronicle the event. A white sports coat and a pink carnation, ah yes, I'm all dressed up for the dance. Cinderella will, of course, wear her special gown, a special 'do, and perhaps even glass slippers to the ball. Memories will last a lifetime, the dance hours.

Years pass as memory fades. Who was my date to the prom? Can't recall? Maybe smoke got in my eyes, or was it the twelfth of never. Then again, maybe someone else caught my eye.

At a noon dance, Lee met Diana Bodia. When she introduced herself, the unexpected invasion into his life was unforgettable, impossible to describe in any detail. He could only surmise the significance of the initial meeting was substantial. No need to ask why it happened because the occurrence was so captivating. The only reality was the inner glow he experienced. Should he pinch himself to insure he wasn't dreaming?

Until the moment of the meeting, he knew absolutely nothing about her. It wouldn't have surprised him to learn she had recently arrived from some distant planet. More probable, the mysterious pretty had just transferred from another school. Whatever, her status was an unknown, a closed book. She was different, alien to most girls he knew or had known. Her independent behavior hypnotized him.

Making his way off the dance floor after dancing with Susan, Lee dodged and twisted through a maze of humanity, attempting to avoid a collision with classmates oblivious to anything outside their province.

A mellow demanding tone penetrated his consciousness, demanding, "I believe an introduction is called for, don't you?"

In stark contrast to the soft melodious sounds coming from the background, the persuasive alto tone of her voice was acrimonious, but like the background melody, its appeal was compelling and enticing.

He turned to determine the identity of the speaker, immediately noting the diminutive form of some unknown, exquisite goddess.

A resplendent beauty, she presented unblemished attractiveness. Freckles excepted, she radiated elegance, but even those small precipitation's of pigment, symmetrically placed ever so carefully upon her tanned skin, seemed to accent her splendor. She exuded a confidence, seemingly able to leap outward without announcement. When she smiled, her eyes flashed. Vulnerability, if she possessed any, lie hidden behind a shield, a sarcastic aura radiating from her essence. Had she possessed membership in the equine family, she would have displayed the temperament of a skittish filly whose bloodline defied predictability. Taming, possessing the spirited coquette risked breaking her spirit. She needed freedom. When first encountered, she appeared almost domesticated, approaching as though seeking an orchestrated form of affection.

Would he ever know whether it was attention she required, or if the opening quest merely represented a tantalizing game not unlike cat and mouse.

"Aren't you going to say hello? The way you're looking at me makes me uncomfortable."

He immediately became party to a completely controlled encounter. The circumstances and the dynamics of the meeting with Diana caused all of his evaluation tools to freeze. Assessment, if it existed, was purely instinctive. He was immediately transformed into a mushy, gooey possessing no power to function independently. He felt like a marionette, she the puppeteer.

She smiled with the smug air of an enchantress empowered by some deity while he continued to stand helplessly before her. She had captivated him, transforming him into a speechless, mummified shape.

"Well, I guess I have to go first. My name is Diana Bodia."

She smiled, almost smirked, as delight appeared to dance in her dark brown eyes.

"Come on! Aren't you going to tell me your name? Maybe you want me to guess."

A warm sensation swept over him. Emanating from the base of his neck, the hot flash quickly enveloped him. Almost immediately his cheeks became a rosy red color and a swarm of butterflies started to flutter in the pit of his stomach.

"Hi."

"Oh, good! He speaks."

Her smile, now a smirk, spread quickly across her face. It indicated the silence should not linger.

"I, I'm Lee Grady."

She shook her head, a first admonishment.

"That is an incorrect indicator. Your name is Lee Grady but you are something else. Oh, I wouldn't worry about it, right now. I'll find out what you are later." Defiantly placing her hands defiantly on her hips, she continued, "You're not leaving yet, are you?" We haven't had our dance."

Her innocent interrogative merely disguised her true objective, the response she apparently expected. Pleased by the placement of her serve, she must have been overjoyed with her subsequent responses to his returns, continuing to keep the question-and-answer volley coming like the German gunners in the Normandy pillboxes.

He was confused, certainly frustrated. He might have even been flattered. *How in the hell does she know me? I've never even seen her before.*

"You dance, don't you?"

He nodded, obediently responding to the petite hand offered.

"Sure."

Seconds later they were dancing, his mind adrift, perhaps marooned.

Responding to his lead, she offered no resistance when the crush of the crowded dance floor forced the distance between them to close. As their cheeks brushed, the force from their closing space brought nearly every critical point of their bodies into contact. Romantic refrains of "The Twelfth of Never" played in the background as he momentarily lost awareness of anyone or anything around him except for an exciting form, suddenly so close, so extraordinarily near.

Invading his dreamy thoughts, she petitioned, "Would you like to walk me to my afternoon class when our song is done?"

"Um, sure. Yeah, I'd like to very much, but, ah, this isn't my song. I like it, but 'Smoke Gets in Your Eyes' is my favorite."

"Really! I like it, too. Um, we could be unique and have two songs."

"Yeah, I suppose."

Man, she's a really good dancer. She makes me feel like Fred Astaire, and she could easily pass for Ginger Rogers.

"You're a great dancer. Have you taken lessons?"

"I'm a student instructor at the Arthur Murray Dance Studio. I learned to dance there by watching the instructors. Um, I'll return the compliment. You're very good yourself. I'll bet you've taken lessons."

"I took a few lessons when I lived in Bend."

"The instruction, was it private or did you have group lessons?"

"A group of mothers purchased a series of lessons from a professional. I guess there were about twenty of us in the class. Our mothers were afraid we'd be wall flowers at the school dances if we didn't learn some of the more common dance steps."

"How nice of the mothers. How would you like to come down to the studio where I work, sometime? If I talk to her right, maybe I can get you hooked up with an instructor by the name of Julie. I'm sure she'd be more than willing to give you a refresher course. Ah, not that you need it, but she's really good, and you'd be able to add a few new steps."

"Such as?"

"Um, the samba, tango, whatever else she has time to teach you."

On their way to class, they passed by the main office.

"I'm gonna pay a visit to the Dean of Women's office to check out the guest book for the prom. Would you mind waiting for me?"

Leaving him midst the turmoil of passing throng, she began muscling her way towards the office.

Suddenly she turned and called to him over the din of the passing students, "I promise, I won't be long."

He'd only known her for ten minutes, but she already occupied his every thought. He had a fever; his heartbeat was irregular. An unexplained fire was racing around in his gut like an anxious pointer searching for a pheasant in an open field dotted with large clumps of tall grass. Debating the possibilities of taking her to the prom, a sickening, hollow sensation crept over him, for without reason he knew he did not want to share her with anyone.

I've gotta get to know her better. I wonder, nah, she must already have a date to the prom. She probably went in to register her date. Why else would she be checking out the prom guest book? Anyway, why should I care? I just met her. I know nothing about her. Why would I even think about asking her to the prom? She wouldn't be interested in going with me anyway. She's not gonna go with someone she hardly knows. I'll probably take Mel's advice and ask Linda Martin to the Prom. She's damned good looking, lots of fun, and, well, we've hit it off pretty good, so far. On the other hand, Diana Bodia would be the perfect prom date, ah, no, what am I thinking about. We've known each other for less than an hour. No way she'd even consider going with me.

In the dean's office, Diana was on a mission. Her search was quick but intensely thorough. Her forefinger stopped at the bottom of the page. She smiled, nodding as she closed the book.

Well, what do you know? No date yet.

Diana's visit to the office claimed less than two minutes. When she returned, her mood was buoyant, almost giddy. Taking his arm as they made their way down the hall towards class, she squeezed it affectionately.

"Want to meet me after school? You could walk me home after school if you don't have anything better to do."

"Really? Sure, I'll see you after school."

He didn't have much to go on when he launched a search for information about her, visiting places where scuttlebutt usually surfaced. The first source, the locker-room gossip chamber, failed to turn up anything substantive about her past or her present activities. Her closest friends weren't much help, either. He was only able to learn she was popular, but nobody really knew her. Other than learning she belonged to the Pep Club, served as the associate editor of the school yearbook, and lived with her aunt and uncle, nothing specific surfaced about the petite, fiery junior brunette. Her biography seemed to be filled with empty pages, especially when it came to uncovering personal information. Was she going with anyone? Whom has she gone out with? All queries evoked no more than a shake of the head or a shrug of the shoulders. Other information was just as vague. Why wasn't more known about her? Over fifteen hundred students in the high school's student body might have been a possible explanation. Then again, maybe her preference for anonymity was the real reason.

She seemed to be a total enigma. From what he could learn, she apparently enjoyed operating in the shadows of student life, a mystification to most. The prospect of someone solving the riddle of her existence probably did not affect her one way or the other. Like the main character of a suspenseful mystery, she presumably would never have sought public notoriety. Publicity derived from an inquiry would probably have affected the questioner far more. Any personal interest in her or her activities probably would require a well designed plan if anyone hoped to lift the curtain of secrecy surrounding her.

When he met Diana at her locker after school, she momentarily let down her guard, uncharacteristically displaying genuine excitement and warmth.

"I'm so glad you agreed to walk me home. Usually I go to the yearbook office for an hour or so, but today I would rather be with you. Ah, do you have to be anywhere special?"

"Not really, at least, not until dinner time. I think my mom would be a little upset if I was late. How about you?"

Diana smiled, motioning towards the exit.

"I start getting ready for work about a quarter to five, five at the latest."

"What time do you have to be at work?"

"I work every night after school from six until nine."

"And weekends?"

"I don't work on weekends."

During the short walk to her home, two blocks south of the school on Fourteenth Street, he failed to learn anything new about her as she playfully darted in and out of one informational trap after another.

She's an artist, very proficient at protecting her privacy. She only reveals what she wants me to know. Even that's hard to come by. Hmm, I wonder if I'll ever know any more about her than I do now? Her ability to guard against unwanted intrusions is impressive, but how in the hell does she ever expect someone to get close?

"Do you mind if I ask why you went into the office to check out the prom guest book?"

"No, but whether I tell you or not is another question." She chuckled, elusively adding, "Let's just say I was checking up on someone."

He'd come tantalizing close to discovering her purpose for visiting the dean's office, but unlike the game of horseshoes, no points were awarded being close. She was a clever, evasive tease.

It's not even realistic for me to expect her to go to the prom with someone she's only known for a few hours. The prom's the most important social function of the year. Maybe before considering the prom, I need to find out if she'll even go out with me.

Stirred to action, realizing the opportunity was rapidly disappearing as they neared her home, he asked, "Would you like to go out sometime?"

"I don't date frequently. My job, school work, and serving as associative editor of the yearbook, ah, they keep me pretty busy."

Undaunted and committed to staying with the provocation until the bitter end, he pressed forward, "I guess that was a no? Ah, do you mean you can't find the time or won't?"

Chuckling, she petitioned, "There's a difference? I just hate it when a person makes an assumption, or asks two questions at the same time. Even if the assumption is incorrect, I'm forced to determine which question deserves to be answered first. Then again, maybe only one of the questions deserves an answer."

"Wow, all that in less than five seconds. You know? I hate it when a person answers a question with a question and follows it with a lecture."

She nodded, appreciative of the resistance. Refusing to raise the curtain of silence, obviously intent upon postponing her response, she was careful not to completely discourage his unrefined courting gesture. Gently touching his arm, she sought communion with his eyes.

"Ah, well, I guess I'd better be going. It's getting near dinnertime. If you don't mind waiting until I get off work, I'd love to go out for a Coke with you some time. It's best we begin our new friendship by easing into the dating routine, don't you think?"

His face reddened, a warm, eerie, glowing sensation taking root at his core and then radiating outwardly.

"I guess that was a yes? You haven't said anything about the weekends, you know, Saturdays. Are you seeing someone?"

"There you go. What was that, three questions?"

She paused to tease his emotions, appearing irritated at the invasion.

Suddenly turning as she started towards the front door of her home, she tossed a sarcastic inquiry over her shoulder, "Whatever would make you ask?"

After a few deliberate, haughty paces, she stopped.

"Um, I guess if you'd like, you can call me later. I should be home by nine-thirty. We discuss it then."

"I could come by and pick you up for a Coke."

"You could, but you won't because I won't let you. Why don't you quit while you're ahead? I see no reason for the rush to get to know each other. If you want, call me at nine-thirty."

When he called later that evening, Diana's aunt answered the phone.

"Hello? This is the Omara residence."

"Is Diana home?"

"May I ask who's calling?"

"This is Lee Grady."

"Um, she's not, ah, she just walked in."

A hint of disinterest dominated Diana's response when she finally answered the phone, "Yes?"

"Um, you suggested I call."

"I did? Well, I guess congratulations are in order. Looks like you accomplished your mission."

"Don't you remember telling me to call after you got off work?"

"Do you always do what you're told?"

"I thought you wanted me to call."

"I guess I could ask you again if you always do what you're told to do, but the truth is I do vaguely remember something about asking you to call. Anyway, I suppose I'm glad you did."

"Me, too. I'd kinda like to finish our discussion."

"You mean about going out?"

She yawned audibly. The ensuing pause was deafening.

"Yeah, that's what you suggested we do when I walked you home from school."

"Hmm, so, I guess you want to know if I'll go out with you on Saturday?"

"Well?"

"I'm considering it. You should know I have no plans to date steadily now or anytime soon, and I only go out on weekends if the right person asks. Um, if you want to know if I'll go out with you Saturday, you might consider being more specific. Why don't you tell me which Saturday you want to go out? Knowing what you have planned to do might also help."

"And you talk about me asking more than one question at a time. Um, anyway, I thought I told you I wanted to go out this Saturday."

"First off, those weren't questions. Next, you might tell me what you have in mind."

"Um, I was thinking about a movie; how does that sound?"

"A movie. Hmm, and you did say Saturday, didn't you?"

"Right. Ah, may I ask why you don't date steadily?"

"You may if you like."

Suddenly the phone seemed lifeless.

After a maddening length of time, she continued, "Nothing against you, Lee, but I have far too much to accomplish to let something as juvenile as going steady get in my way. I still have my senior year ahead of me."

"I never said anything about going steady, but since you keep bringing it up, have you ever gone with anybody?"

"Persistent aren't you? Hmm, I guess it won't hurt to tell you I recently broke off a relationship with a really neat guy from Willamette University. He played on the basketball team and was studying to become a doctor."

"If it's any of my business, may I ask why?"

"If you're asking why we broke it off, let's just say, both of have a lot to accomplish. Ah, we agreed to break it off before either of us got hurt."

"Oh."

"Yeah, I plan to stay unattached during my senior year and begin college without any commitments. Still interested in asking me out?"

Lee's dating resumé was not extensive, documented experience nonexistent. His association with Susan Sloan, a brief encounter, lasted one date. Mary Ann only encouraged casual flirting. Linda Martin was receptive, but too easy to take advantage of, and the encounter he had experienced with Sandy was only a possible bookmark for the future. No wonder he was apprehensive when he considered initiating a relationship with Diana. He feared a serious dating gesture would surely fail, but she had also awakened strong feelings. He was hooked. He wanted to escort her to the prom, for that matter, anywhere. The conditions didn't matter.

"Yes. How about taking in a movie with me this Saturday?"

A seemingly endless break in their communication again began. Watching the sand slowly fall from the top of the hourglass, the pause probably was intended to surpass the time it took him to respond to her inquiry.

Finally, breaking the silence, she offered, "I'll let you know tomorrow. Maybe we can do lunch."

She yawned again.

"Excuse me! It's past my bedtime. See you tomorrow."

The contrast between them was quite noticeable, his manner resembling an open book. Any strategy, if it existed, appeared in bold print, the language simple. Her copy was hard to read, similar to reading three-dimensional print without special glasses. Going sailing with him would have been easy. The only necessity, a carefully plotted course and a little caution to keep the ship from going aground. Responses from her were very unpredictable, variable. Aloof, at times mysterious, Diana even could be genuinely warm. Whatever personality she put forth, there was something about her that fueled an overpowering interest, making the blood run hot through his veins. He couldn't resist the attraction he felt for her even though her mannerisms confused and discouraged him from opening up to her. Down deep, he knew instability and unforeseen encounters lie ahead if he were to win her heart.

Diana's refusal to even commit to a Saturday night engagement irritated him.

Mel has been lobbying for me to take Linda to the prom for the past three weeks. I'm guess I'll call and ask her to the prom. At least she wants to go out with me.

The attractive South Salem High School junior was a bubbling fountain of spontaneity. Her rich bronze-toned skin was framed by long shoulder-length blonde hair. The first time he took her out, the date ended with her initiating a long passionate goodnight kiss. Linda's direct and warm demeanor flattered him, but she was an easy target. He often called her for a date at the last moment. She had accepted his casual advances, responding eagerly when he fondled her breasts or briefly explored her more sensitive regions.

It's easy to tell when I've gone far enough. She pushes my hand away or clears her throat when she doesn't want to go any further. She never makes an issue of it, but it's pretty clear I could go further if I would ask her to go steady. Damn, I'm not sure that's what I want.

He couldn't put his finger on it until he met Diana, but Linda lacked an air of intrigue and mystery. Reading her diary didn't even require opening the cover.

"Hello?"

"Hi, Linda, this is Lee. I hope I didn't call too late."

"Don't be silly, I don't go to bed until ten."

He checked the clock in the kitchen and chuckled.

"Good, we still have ten minutes. Um, how are you?"

"Fine. Ah, I've been thinking about you. I'm so glad you called."

"I've been thinking about you, too. Um, I called to find out if you'd like to go to prom."

"I'd love to. I was thinking about asking you to my prom, but when I found out our proms are the same night, well, I was kinda keeping my fingers crossed you'd ask me to yours. Are we going with Mel and Shelly?"

"I haven't talked with Mel about it, but I'm sure we will. That's okay, isn't it?"

"Of course. Shelly and I are becoming great friends."

Lee yawned.

"Excuse me! I need to head for bed. Um, I'll call you tomorrow. Maybe we can go out for a Coke or something."

"If you do call for a Coke date, we can't stay out late. I have a big test in history."

"I'll call you about six. We can talk about the prom then."

It wasn't unusual for him not to see Diana between classes in the morning. Failing to connect with her at lunch, was cause for concern. At the end of the school day, another quick drop-by visit to her locker came up short.

We were supposed to have lunch together. I wonder if she's sick? Let me see. Maybe I should check out the yearbook office. If she's not there, I guess I'll just have to give her a call.

When he entered the yearbook office, steely glares greeted him.

Hmm, friendly in here. You'd think someone died or something.

"Have you seen Diana?"

"She left a moment ago. She said something about going home. Funny, she usually spends at least an hour or so in here every night after school. Come to think about it, she seemed upset about something. I sure hope everything's okay."

"Me, too. Um, thanks, I'll see if I can catch her before she gets home."

From a distance, he could tell she was in a hurry, carrying herself like a wounded animal. Although retreating, her mannerisms indicated an encounter could be perilous. Eagerness, however, overruled common sense.

"Diana, wait up a minute! I've been looking for you all day."

At the knee-high brick wall in front of her home, she slowed her pace, and then dramatically stopped. Wheeling about, she defiantly glared at him. Visibly angry, the manufactured smile on her face, a vindictive sneer, indicated he was about to confront a dangerous antagonist.

"Aren't you lucky? You finally found me."

Her eyes flashed.

"What do you want? I have to get ready for work."

"I thought you didn't go in until six."

"Tonight, I have to go to work early."

"I looked for you all day. I even tried finding you at your locker and in the yearbook office."

"You've been busy."

"Don't you remember? Last night you suggested we have lunch."

"I did? How rude of me to forget, oh, wait a minute. I do faintly recall suggesting that maybe we could do lunch."

"Have you made a decision about Saturday night?"

Her eyes started to redden, tears threatening to appear.

"Yes. Um, I've decided not to go."

"Oh! I see."

He ducked his head.

"I doubt that."

"I guess I misunderstood."

"Obviously you did!"

Methodically she set her books on the top of the brick wall surrounding the front entry to her home and then defiantly placed her hands on her hips. Her emotions now under control, she faced her suitor with overpowering self-assurance.

"Did you ever stop to think when I failed to meet you for lunch I was trying to give you a hint?"

"I guess I've really made a mess of things."

"Really?"

"I wanted to ask you to the prom, but I didn't think you would go with me. Um, I asked someone else."

"You're good at making assumptions, aren't you? I suppose I should be flattered you considered asking me, but, in truth, I could care less."

He shrugged, realizing the futility of saying anything further.

"Humph. Just like you to lose your tongue at a time like this. Anyway, we've known each other less than two days, too little time to even consider something as important as going to the prom together. I would think you would want to know the person you take."

"I suppose you're right."

Her eyes were glassy black pools of seething anger. Her voice was no longer steady.

"Besides, I've already been asked."

"Oh. Now I really do feel stupid."

Shrugging, she offered, "We wouldn't make a very good match, anyway."

In the distance he could the sound of a whistle as the four o'clock train crept through the congestion of Salem's late afternoon traffic. If it had played chess, the sound would have indicated it was check, perhaps even checkmate.

"I guess I'd better go. I'm sorry about the mess I've created. He sighed, "I won't bother you any more."

He turned, and started in the direction of the Viking campus.

"If you'd like, we could try it again, start all over. Why don't you come by my locker before lunch tomorrow? We could do lunch."

"You suggested we have lunch today when I called you last night."

"I know. Today has been, ah, well, I've been in a foul mood all day."

"Right now, I think I know how you feel. Um, I'll come by your locker tomorrow. If you're there, we'll do lunch." Shrugging with resignation, he submitted, "Otherwise, I guess I'll see you when I see you."

"How profound." Grinning, she mandated, "You'd better wait if I'm not there because I fully intend to keep the appointment. We have a few things to discuss."

He did not enjoy the walk home. Thoughts about whether he should bother meeting her swirled through his head.

I love a challenge, but she's way beyond that. She's impossible. I'm glad I'm taking Linda to the prom. It'll be nice being with someone who doesn't work overtime trying to make me miserable.

Lee and Diana shared few traits. One they did share was their proficient ability to disguise their feelings. For whatever reason, they refused be open with each other. Diana opted to hide behind a mask of aloofness, content to play hard to get. Lee opted to hide his feelings in a closet of insecurity with a lack of confidence hanging nearby.

The next day, Mel ran into Lee just before lunch.

"I talked to Shelly and she's all for doubling to the prom. Ah, I suppose you've heard that Diana's going to the prom with Jack Pram. I guess he's a real lady's man. Somehow I don't see them being real compatible."

"What difference does it make? I really like her, but I guess she just doesn't feel the same way."

"She doesn't let anyone get close. Besides, you've only know her for a little while."

"It doesn't stop me from feeling the way I do."

"Maybe so, but if I were you I'd throttle it back. I sure a hell wouldn't caught overboard without a lifeline. She works too hard covering up who she really is to be good for you."

"You may be right. She does seem to have a powerful need to be something she isn't, yet so unhappy with the person she's become. Damn, I still can't help feeling the way I do about her." Grinning, he confessed, "You know? I even like her unpredictability."

"Have you told her how you feel?"

"Are you kidding? That would be suicide."

"Why?"

"She takes such delight in making people around her pay the price for her unhappiness. I don't know why she works so hard to make my life miserable other than maybe she's afraid of letting her guard down."

"Maybe she's afraid to expose herself. You're no different, you know. You can't even bring yourself to tell her how you feel."

"We'll finish this later, Mel. Diana's coming. Ah, we're supposed to do lunch."

Smiling as she approached, Diana innocently inquired, "Am I late?"

"No, you're right on time. I'm glad to see you."

"I'm glad to see you, too. Ah, I wasn't sure you'd be here."

"Isn't that what I was supposed to say?"

"I'm sorry I was so short yesterday. Nothing seemed to go right."

"I really wanted to ask you to the prom."

"That's one topic I don't care to discuss. I have no intentions of crying over a little spilled milk."

"Sorry, I though you would like to know why I didn't ask you."

Unexpectedly she exposed a hint of her frailty, tears welling up in her dark eyes. As quickly as her inner feelings had been uncovered, her steely exterior reappeared, the all too familiar protective mask of indifference again greeting him.

"It's for the best anyway. You and I are like water and hot oil. Maybe we should quit while we're ahead."

"Last night I might have agreed with you, but now…."

"You're unbelievable! Other than a dance at noon and two or three meetings, what could we possibly have going?"

"Maybe if you were willing to give us a chance, we'd…."

"The pronoun 'us' denotes togetherness and involvement. It calls for a commitment. I not ready to make promises."

"Okay, so why don't you tell me what you want instead of sparring with me all the time?"

"Being friends would be okay."

"Okay? By your definition, becoming friends means not getting involved.

If we don't go out with each other, we should become great friends. Wow! Now all we have to do is decide how much space we give each other so you can really be happy."

"You missed my point."

"Well, just what is your point?"

"I wouldn't mind going out with you as long as we keep it casual. Ah, I have no intention of allowing it to become anything more. By the way, I'd love to go to a movie this Saturday if you still want to take me."

"Sounds good, but I want something understood."

Grinning, she inquired, "Okay, tough guy, what do you have in mind?"

"If I agree to keep everything casual, then I want you to stop mentioning you don't want to get involved. Can't we just let things evolve? I promise not to push, but you have to stop setting up barriers."

"Hmm, I'm not used to someone else designing the rules. Oh well, I guess the only way I'm going to get you to stop pressing the issue all the time is for me to stop being so difficult."

The prom made an impression for reasons other than Lee had hoped. The North Salem gym was beautifully decorated and the band played all of his favorite songs, but something was missing. "Oh Diana," summed up the frustration he felt about the evening. "A White Sports Coat" gave ample evidence of how hard he had tried connecting with Diana, but "The Twelfth of Never" and "Smoke Gets in Your Eyes" were constant reminders that he was with the wrong girl. Absolutely breathtaking, Linda did everything to make his senior prom a night to remember. Not his finest hour, he fell far short of showing her the appreciation she deserved.

On the other side of the floor, Diana radiated her usual beauty, but even thought she hid behind her all-too-familiar protective shield, she couldn't hide her disappointment. When a last rendition pleaded for someone to save the last dance, they weren't able to share the enjoyment with each other. Instead, they were forced be consoled with a few stolen glances and the satisfaction of knowing two couples shared the dance floor but not each other. When he kissed Linda goodnight, he knew it was far more. Their brief relationship ended as their lips parted.

Mel and Lee stopped by the Uptown Drive-in after taking their date's home.

"Shelly and Linda wanted to go to the coast, you know?"

"I wasn't up to it. Besides, I have to go to work early tomorrow. Why didn't you take Shelly?"

"We didn't want to go by ourselves. Actually, I wasn't up to it, either. Shelly's a good friend, but she's not the one I really wanted to take. I guess you had the same problem, huh?"

"Linda's great. I just don't have strong feelings for her."

"Sounds familiar. I'm trying to work things out with Jackie."

"Now you know how I feel about Diana."

"You were thinking about her all night, weren't you?"

"I can't seem to get her out of my mind. Ever since I first met her, she has done something to me I can't explain. It's almost like I'm possessed. Ah, I know you feel I should have treated her differently, but I wanted to be with Diana so bad."

"I don't think Diana's right for you. She doesn't even compare to Sandy, or have you already forgotten about her?"

"No, I haven't forgotten about her. It's just…. Ah, what makes you say Diana isn't right for me?"

"You're both like a door somebody has closed and bolted shut. The only difference is you're trying to unlock it. I wish I could convince you to leave her alone. I'd hate to see you get hurt."

"I appreciate how you feel, but I'm still gonna give it a shot. Um, it's one o'clock. This place is getting ready to close."

"And you do have to go to work early. Even though I can't convince you Diana's wrong for you, I'd like to suggest you still keep your thoughts about Sandy alive. Of all the women you've known, she's, well, she's crème de la crème."

"Yeah, suppose I met the right girl at the wrong time?"

"Maybe and maybe not. Didn't she say something about hooking up down the road?"

"Yeah, but somehow I feel the road is long and winding. Sometimes I wonder if the trip would be worth the effort."

Pulling out of the Uptown Drive-in, Mel grinned. Suddenly, silent humor escalated to a more audible chuckle.

"What?"

Giving his friend a friendly nudge on the shoulder, Mel offered, "Trust me, Lee. The effort will be worth it. I just hope you have the patience to stay the course."

PHI ALPHA, BROTHERHOOD, AND MINERVA, TOO

From the originating point in life, parents, the extended family, the neighborhood, and other societal influences strive to mold a person's behavioral patterns. Formal schooling kicks in near the fifth year, transmuting subscribers, willing or unwilling, into reasonably functional, literate contributors to society. Twelve years, perhaps thirteen years of schooling ends with a formal ceremony proclaiming the eighteen-year sentence served. Nervous human configurations gingerly toe the first step of the platform. Almost hesitantly they mount the stage where other commencement participants previously passed as amplifiers broadcast legal names and surnames to an audience of proud family members and friends.

Emancipation! How rapidly the years passed as boyhood merged into manhood, the little girl became a young woman. Magically, the graduates begin a journey into the future. The birds soar skyward to try their wings.

The venture beyond the learning chambers of North Salem High School ushered in a unique twelve-month span, four distinct seasons. Residents new to the Willamette Valley, normally used to experiencing diverse seasonal pronouncements, hardly took notice. Long time residents, however, reveled in the rarity kicked off by a hot summer.

Lee's schedule rapidly became a whirlwind of hectic activity. A bustling, complex schedule fraught with unpredictable hours he had to work at Borg's was jammed and difficult to coordinate.

"I got an invitation to attend a fraternity function."

"I hope you're going to accept."

"I was thinking about it, but it really puts me in a bind."

"How so?"

"If I go, I won't be able to see you this weekend. Damn, trying to juggle my schedule at Borg's, take you out, and find time to visit with friends going away to college is touch enough without...."

"You have to do what you have to do. Ah, I start working at Blue Lake Packers this weekend. We're going to have even less time together."

"You still gonna work at Arthur Murray's?"

"Why not? It's a great job. Besides, the money's too good to pass up."

"What hours will you be working in the cannery?"

"Nights. I'll start at eleven and get off at seven."

Smirking, she suggested, "We'll still be able to go out."

"When?"

"I get off at nine at the dance studio."

"Wow! Two whole hours."

A visit to Paradise Island's entertainment park lost its luster, an uncomfortable discussion fueling a diminishing glow to what had started out to be a very idyllic evening.

"When you start school at Willamette this fall, I want you to start dating other people. I still have no intention of letting our involvement stand in the way of my senior year's activities." Smiling, she continued, "This is a going to be a special year for both of us. We both need the freedom to date other people."

"But I don't want to go out with anyone else."

"I promise not to get jealous." Glaring at him contemptuously, she again elucidated, "Besides, it's not open to negotiation. I'm not gonna tie myself down until I'm in college."

"I don't see what it'll hurt if we continue to go out with each other."

"Come on, Lee. Stop being so difficult! I wasn't suggesting we stop seeing each other all together."

"I don't know. This doesn't sound–"

"I doubt it'll be the ruination of your life. My feelings for you aren't gonna change just because we go out with other people."

Nearly four months had not strengthened his position with her one iota, her domination of the relationship undeniable. The best he could expect was occasional displays of guarded fondness, a light touch on the arm or a gentle squeeze of his hand. Even during rare romantic moments, she found a way to redirect his impulses. Still, he had an obsession.

Residents in Salem sweltered midst uncomfortable August temperatures as the sky began to display a depressing blandness, subtly announcing summer's retreat. Cool evenings encouraged patio dwellers to put on sweaters to enjoy an outdoor grilling session, as, almost simultaneously, the harvest moon started making nightly appearances. Hanging tantalizing close to the horizon, its arrival seemed to coincide with the gradual evaporation of daylight hours.

With the signals of fall's arrival, many of Lee's friends began preparing for life after high school. After seeing Mark and Marty Lee off for school in Montana, he went by to see Mel.

It's not gonna be easy to say goodbye to Mel. It won't be the same with him down at the U of O. We've been friends since I first came to Salem. Besides, who else am I gonna have to share problems with my love life?

"Didn't get enough of the Oregon campus during the basketball tournament, huh?"

Grinning, Mel revealed, "My uncle's going to let me work part-time for him to help with the expenses."

"We're only a few miles apart, so don't be a stranger."

"It's a two-way street. I'm sure the next four years are going to be a busy time for both of us, but there'll always be holidays and school breaks. Ah, don't tie yourself down with Diana, Lee. Give the coeds at Willamette a chance."

The seasonal transition from summer to winter in the Willamette Valley was usually a mere passage of time, autumn passing through like a disinterested tourist. This year, the seasonal transition took time to record the moment with significant gestures. The consequence was a spectacle of beauty. Snowing mountainous drifts of bright multi-colored leafy wonderment upon the terrain, the alluvium gatherings signaled an end to a season of nurturing and growth as unenthusiastic tenants took rakes and other leaf gathering equipment out of storage.

As the remains disappeared efficiently into conveyances, Lee plunged into freshman orientation, registering for classes, purchasing books and paying fees. Rush week followed. His exposure to fraternity life had not impressed him, but the Greek organizations at Willamette somehow seemed different.

He accepted an offer from the Sigma Alpha Epsilon fraternity.

The next evening, proudly displaying his pledge, he revealed, "I decided to pledge, after all, Diana."

"My 'ex' was a member of the Beta Theta Pi house, but he almost went SAE."

Caressing his pledge pin, she smiled devilishly as she slipped it under her blouse.

"Do you know what you have to do if you decide to offer this pin to someone before becoming a member?"

"I wasn't aware you'd be interested in such nonsense."

Smiling coyly, she persisted, "Don't you even want to know what you have to do?"

"Okay, what would I have to do?"

"In the presence of a member, you have to pin this little badge in a very special place."

"Uh-huh! Not somewhere visible I suspect."

Giggling, she continued, "You'd have to pin it under your girl's bra."

His face immediately assumed a crimson glow.

"Yeah, right. You'd knock down barriers to be a part of that ceremony, wouldn't you?"

"In your dreams! Well, now that you've picked a fraternity, it's time to hit the books so you can make grades for initiation."

"I know. Classes begin Monday."

The Bearcat campus had an ambiance similar to Ivy League schools, featuring carefully shaped hedges, beautiful flowering shrubs and a well-manicured lawn crisscrossed by a highway of cement walkways. The buildings sported growths of ivy, seemingly intent to reach heaven by way of the brick or stone structures acting as their trellis.

Small by most standards, the school's campus had breathtaking splendor throughout its continuous configuration of ten square blocks. A meandering stream, Mill Creek and tall stately deciduous trees separated the undeveloped southern half from the northern part of the campus. State and Twelfth Streets formed the northern and eastern borders. Resting midst state government offices, the Thomas Kay Woolen Mills, the city library and the tall spired Methodist Church, Willamette's quiet campus appeared to exist in isolation. Only the congestion created by the five o'clock rush hour or the noise from a passing train disrupted the serenity of the campus.

At first the peaceful academic environment mesmerized him, and then sounded a loud warning bell. When he roused to the peal of the midterm alarm signal, his boat had capsized in the rough academic seas.

Damn, I've never had below an 'A' in a math class before in my life. This 'C' is embarrassing. Hmm, it's not too surprising what the history prof has to say.

A sheepish smile spread slowly across his face.

Looks like my midterm essay lacked organization and documentation. I guess all things considered, a 'C' isn't too bad. You have to be kidding, a 'C' in ROTC? Holy shit, I flunked foreign language. Guess I didn't inherit Grandma's ability to sprechen sie Deutsche.

Disappointment consumed him as his cheeks assumed a crimson glow and tears began to fill his eyes.

English is such a waste of time. I hated it in high school and it isn't a damn better here. Even bonehead English causes me problems.

He stuffed the report into his pocket.

Fuck! I either bring up my grades by the end of the semester or I'm history. I don't understand how I could get 'Bs' in high school by hardly cracking a book. I sure didn't expect college to be this tough. I suppose it might help if I eliminate some of the fraternity bullshit and hit the books a little sooner than the night before a test. This is a great night to have a date with Diana. Wait till she finds out about my grades.

After taking in a movie, Lee parked his parent's '57 Chevy Impala in front of her home. The climate was chilly.

"You've hardly said a word tonight. Sometimes when I'm with you, I feel like I'm wasting my time."

"Maybe yes, and maybe no. It all depends on how seriously you decide to take getting a college education. Speaking of which, ah, how did your midterms turn out?"

Shrugging, he confessed, "I could have done better."

"Care to be more specific?"

"I almost got a two-point."

"Almost? Hopefully your GPA was above that modest standard."

"Ah, no. It was a little below."

"Hmm, which course was your Waterloo?"

"German. Ah, I flunked it."

"Uh-huh! And the rest of your courses?"

"I got 'Cs'."

"Even in math? You must be so proud." Shaking her head, she counseled, "It's really too bad because you're bright enough to do so much better." She glared at him for a moment, again shaking her head in disgust. "I don't think our relationship is going anywhere. Ah, I have to go."

Not waiting for him to open the door for her, she exited the car. Suddenly she wheeled around towards him and bent down so she could capture his gaze.

"Just know we won't be going out again until I have evidence you're passing all of your classes with at least a 'C.'. And, ah, don't bother calling. As far as you're concerned, I'm not at home."

After she had disappeared into her home, he sat slumped over the steering wheel.

To hell with her, I'm tired of always dancing to her tune. There's no way I'll ever pass German. Well, I might as well enjoy the time I have left at Willamette. No sense brooding about it.

He slammed his fist into the dash and then slowly reached down and turned the key to the ignition. He gunned the engine two to three times and then jammed his foot to the floorboard. Squealing tires and a sizeable deposit of rubber announced his departure.

See ya around, Diana. It's been nice while it lasted.

With the relationship with Diana on the rocks, he launched into even more involvement with the fraternity and pursuits totally unrelated to improving his grades. Honoring the expression, "Misery loves company," he joined two of his floundering pledge brothers, also victims of the "Midterm Reaper."

Jerry Moe was the perfect model for an Air Force recruitment poster. Nearly six feet tall and two hundred pounds, he exuded poise and military bearing. His closely cropped blonde hair, steel blue eyes and a broad toothy smile attracted people's attention, his outgoing personality clinching the deal. Jerry had delayed going to college for a year so he could earn enough to cover expenses for the first two years at Willamette. Even after enrolling, he continued to work full time and perform weekend gigs in a local band. Like Lee, his girlfriend had given him the cold shoulder.

Jay Thomas had graduated from cross-town rival, South Salem High School. His love of the outdoors was perfectly blended with a captivating personality and handsome dark features. Women were attracted to him like bears to honey. Modest and quiet, his true personable qualities were seemingly hidden behind thick glasses. Only a ruse, when he emerged from the phone booth, the transformed Clark Kent was a dynamo. Never lacking for female companionship, Jay did everything to avoid academic engagement.

They formed an interesting, inseparable trio, frequenting any gathering in Salem's broad social milieu wherever they could find women to fraternize with, looking for any excuse to party.

"Who are you gonna take to the fall house dance, Lee?"

"You know the foxy senior from North I met the other night?"

"You're not serious, are you?"

"Sure, why not?"

"What about Diana?"

"We're history."

"I think you should give her a call. You know she's the one you'd like to take."

"Okay, so I call her. Should I call Jennifer Joluie and cancel the date first or wait to see what Diana says?"

Grinning, Jay advised, "A bird in the hand is worth two in the bush."

The next evening, Diana's familiar voice intoned the usual distant, disinterested response.

"Hello, this is Diana speaking."

"Diana, this is Lee."

"I thought—"

"I know. You told me not to call until I was passing all of my classes. I can't wait. The semester doesn't end until January."

"That's supposed to make a difference?"

"At least give me a moment to explain, will ya?"

"The clock's running."

"My fraternity house dance is this coming weekend."

"What? And you're only giving me four days' notice?"

"I wasn't sure you'd even consider going with me."

"Seems like this tune has played before. Ah, you're still on probation, you know? We shouldn't even be having this conversation."

"I know, but, ah, you're the only person I want to take to the dance."

"You really don't play fair. If I say no, you probably won't go, but if I say yes, I'll be giving you the wrong message. Isn't that called blackmail?"

"I'm just asking you to my house dance. Ah, I know I have no right to expect you to say yes, but I'd really like for you to go with me."

"Uh-huh. I would hate to see you miss it, ah...."

"Are you saying yes?"

"Oh, I suppose. Why is it I get the feeling you knew all along that I would accept? When you throw your irresistible charm at me, well, sometimes you convince me to do things I know I'll come to regret."

"For your information, it would be very disappointing if my grades aren't already starting to improve."

"It would, wouldn't it? Well, thank you for asking me. Call me Friday and fill me in on all the details. In the mean time, I have a lot of things to do to get ready."

In the yearbook office on Friday, Jennifer Joluie peered over Diana's shoulder to look at the class homeroom layouts.

"Wow, those are really good. Um, I heard you're going to the SAE dance with Lee Grady."

Diana nodded, continuing to check if the names and pictures matched in the layout.

"I'd better get a move-on. I have to call for a hair appointment."

"You can have mine. Looks like I won't need it anymore. Um, it's at Phagen's, two o'clock on Saturday."

"Gee, that's really nice of you. Are you sure?"

"Positive. Seems Lee asked both of us to the dance. Yep, when he broke the date with me, he mentioned he really liked you."

"Really. Hmm, are you okay? You've been acting real strange since we got here this afternoon."

"I'm fine." Wrinkling up her nose, she continued, "Thank God I didn't go to the expense of buying a dress. For whatever it's worth, it's really no big deal. I was only going with him because I thought it would be cool to go to a fraternity dance."

"How well do you know him?"

"A friend of mine introduced us a couple of weeks ago at the Uptown Drive-in."

"How nice. Well, I really don't know what to say other than I'm really sorry. And, ah, thanks for allowing me to use your appointment. Should I call about the change?"

"No, I'll take care of everything."

As much as I hate what Lee did, Jennifer probably got what she deserves. The little bitch is notorious for dumping on people. Damn it! Why'd he have to ask her to the house dance? I can't stand Jennifer, but what he did is unforgivable, almost unforgivable. Hmm, I think Mr. Lee's in for a most interesting evening.

Saturday evening, Lee bounded up the stairs to Diana's front door, his excitement bubbling over as he rang the doorbell.

The house dance is gonna be so cool. Maybe after tonight, yeah, after tonight....

A moment later, Diana opened the door.

"Hmm, looks like I forgot to turn on the porch light. Suppose it'll help if I throw a little light on the subject?"

She's absolutely beautiful. The black cocktail dress she's wearing is, ah, wow!

"I didn't even notice."

"There that's better. Now you can see my dress better."

"You look fantastic. Looks like you had your hair done."

"When you told me it was a dinner dance, I thought I should splurge a little. This little number I'm wearing should do the trick, don't ya think?"

"Oh, yeah. Your dress is fabulous. You look like a million bucks."

"Really?"

Smiling coyly as she handed him her coat, she coaxed, "Ah, can't I talk you into a little more?"

"Possibly. Ah, what time do you have to be in?"

Turning towards where her aunt was sitting, she announced, "I'm leaving, Margaret. I'll be home by one o' clock. I'm locking the front door."

Turning back towards Lee, she slipped her arm through his.

"Shall we go? About the hair, I decided a little change was in order. A girl can't be too predictable, can she? Hmm, looks like you went all-out yourself."

"Thanks. Oh, I forgot to tell you, we're doubling with Jay and Sara."

"It wasn't the only thing you forgot to tell me, was it?"

"Ah, what do you mean?"

"We'll discuss it later. Ah, the door. Um, don't get any ideas just because we're gonna occupy the back seat." After they had both stationed themselves in the back seat, she announced, "No, on second thought, while we're on the way to the dance maybe you should tell me about Jennifer."

His face reddened.

"Jennifer, ah, yes, Jennifer."

"You do know whom I'm talking about, right?"

"Ah, yeah. Ah, I wanted to bring you to the house dance, Diana. It wouldn't have been the same with her."

"Uh-huh. Why did you ask her then?"

"Would it help if I told you I was sorry?"

"Sorry, I'm the wrong person for that petition." Shaking her head, she scolded, "I like you very much, but I don't like what you do sometimes. You have all the potential in the world, but you persist in just throwing it away."

"Maybe so. Why don't we change the subject?"

"Look, I'm sorry you're not doing well in school, but I'm not gonna do you any favors if...."

She shook her head, tears starting to fill her eyes.

"You need to start focusing. Don't try to pacify me by telling me you know it, either. I'd like to see some results instead of empty promises. By the way, you're lucky Jennifer didn't go to any trouble getting ready for the dance. She may be my least favorite person, but I'm still not pleased how you handled it."

Grinning, he teased, "Sure you didn't make a list?"

"Are you getting tired of me complaining?"

"I was thinking we might find a more pleasant activity. Um, it's a good thing I don't have anything to complain about."

"Why?"

"Your curfew. It's one o'clock, isn't it?"

Suddenly a stirring rendition of their favorite song began to play.

"Saved by the bell." Smiling as distance between them closed, she purred, "I told my aunt one o'clock, but I have a little leeway if necessary."

"You know how I really feel about you, don't you?"

"Enlighten me."

"I've liked you from the first moment I laid eyes on you."

"Wow! And I didn't even have to give you a prompt." Smirking, she teased, "And I really like you, too."

"But I'm still on probation, huh?"

"Uh-huh. You still have to prove you're serious about your studies."

At the end of the evening, still facing an improbable future, he bent forward to meet her respondent lips. A first embrace, it initiated an electric charge hurrying to race up and down his spine. After their lips had parted, he gazed into her dark, mysterious eyes.

"I wish...."

She put her finger to his lips.

"I know. It's really strange. For the first time I know I'm probably with the person I want to spend the rest of my life with, knowing it probably won't happen."

She looked away to wipe a tear from her eye.

"I hope this isn't goodbye."

Memories of the house dance had faded when, a week before the Willamette students quit the campus for Christmas break, the skies opened. An avalanche of snow was dumped on the residents of Salem, initiating the third season. Although not unusual for the Capital City to get snow, the storm blanketed Salem with a thick, traffic-snarling coat. Nearly forty-eight hours later, Salem was paralyzed.

The university continued to hold classes even though the campus resembled a ghost town. If not for students passing from class to class, evidence of human life would have been nonexistent. Through it all, Lee's presence was a mere formality. The instructors, abstractions to him, continued to impart information in the same monotonous manner as usual, but other than a name on a registration form, he really didn't know any of them. They merely occupied space at the front of the lecture hall.

One of the instructors, his English professor, Mom Bagwell, was different. A beacon in a dark stormy night, she piloted freshmen through English. Always on the go, scurrying about the campus in her tennis shoes, she carried a briefcase full of caring motherly influence. She exhorted all of her students to meet rigid but fair standards, speaking convincingly of adopting a reverence for enlightenment. Her wrinkled, leathery face seemed to shine when discussing her students, serving as their defense attorney. Never dwelling upon a student's deficiencies, she extolled their virtues, encouraging growth toward academic maturity. However, nobody dared to cross the sly, aging mentor. Sharp of tongue, she was a master of delivering vigorous admonitions to students failing to meet her expectations.

It didn't go unnoticed that Lee shunned her course. Anguishing over the mounting total of missed classes, she noted his attendance registry resembled the record of an AWOL soldier. She came calling on a snowy Tuesday evening, three days before Christmas break.

After a private meeting with the house vice president, John Dort, Lee was brought into the conference.

"Lee, the purpose for Mom's visit is to find out if there's anything she can do to rescue you from disaster."

John smiled, looking at the aging mentor affectionately.

"The members of SAE and I call her Mom because of the many members, me included, she's steered back on course."

"Lee, John and I want to know what it's going to take to convince you to put some effort into the English class. Your mid-term grade was a 'C.' Do you have any idea what you have going as we speak?"

"Pretty bad, I suppose."

She nodded, offering, "It's not too late to salvage a 'C.'"

"I've always hated English. It's the grammar, ah, it really turns me off."

A tear crept into the corner of her eye.

"Do you have any idea how well you write?"

"Not really, but I really enjoy…." Smiling sheepishly, he confessed, "I could live without all the red marks, though."

"Uh-huh! Well, I'd hate to see your creative ideas crushed by the stroke of my red pen, so I'll tell you what I'm willing to do."

"I think this is when I have to make a commitment."

"Uh-huh. First you have to start coming to class. I want you to start writing for me, anything you like. I just want you to write. I promise to read your material without making red marks. I'll give you two grades. One will be for content and the other will be for grammatical correctness."

"What's the catch?"

A quarter of an hour later, Mom and John had extracted a commitment. In less time than he would have spent attending class, he and Mom had a covenant. But had it occurred in time? Would it influence how he approached the other courses he was taking or would take in the future? Like a drastic change in the weather, the event could have been pivotal.

PICKING UP THE PIECES

Probation! A university's version of slapping a floundering student into irons. Failure to take advantage of a second chance is cause for dismissal. Inactive! The definition attached to a pledge no longer able to attend fraternity activities. Purgatory! Requiring penance for an opportunity to attain a desired goal or destination. Extrication! A separation, ending an involvement either permanently or temporarily.

As they walked out of the meeting with John Dort, the new house president, Jerry grumbled, "An entire semester down the drain. Fuck! I totally wasted my money coming here. Everything, tuition, books, and fraternity dues are down the tube. I'm thinking about quitting."

"I thought you wanted to be a pilot and fly jets."

"I do, Jay. I've always dreamed of getting a commission in the Air Force, but I'll be damned if I know whether it's worth it anymore. The only thing I know for sure is, ah, I'm not gonna stop working at Meier and Franks just to attend some damn study table."

"I'm not gonna attend study table, either, but, it's stupid to talk about quitting. My dad would disown me if I did. What about you, Lee?"

"I don't know. I suppose knuckling down and getting my grades up would be a start." Grinning sheepishly, he continued, "You suppose we came here to do something besides screw around? I think it's time I started taking this place serious. Attending study table, as bad as it sounds, will probably do me some good."

He sighed, a look of disappointment spreading across his face.

"Yeah, at least I won't have Diana to worry about this next semester." Jay frowned, offering, "I think a brain transplant is my best hope."

The fourth distinct season vaulted onto the scene with flowery splendor. Life began anew as melting pools nourished by icy deposits started appearing. The season of life wasted no time in producing climatic changes, delighting the recreational advocates with a soothing salve for jangled nerves created by a harsh winter. Overnight, the Willamette Valley's majestic flowery grandeur began bathing in warm sunny days and equally pleasant evenings.

With the emergence of the uncommonly warm weather, the calendar of scheduled campus activities turned the page to an event steeped in tradition, Freshman Glee. An annual affair, it had been held almost as long as the university had been in existence. Glee was a competition between the classes, requiring them to compose a song and choreograph a formation depicting a designated theme. Worthy of special attention and unbelievable preparation, Glee chiseled an entire week out of an already crowded schedule by sandwiching a week of marching and singing practices into the routine of attending classes. The preparation for Saturday evening's performance created tensions and fatigue but when all the classes presented their interpretations, all participants agreed the sacrifice was worth it. The audience must have agreed because as each of the four songs filled the air in the ancient Willamette gymnasium, and four classes marched with pride and precision to form an intricate configuration of prairie schooners crossing the plains, the response was overwhelming.

At the first Glee practice, after learning the administration had lifted his probation to participate in Glee, Lee and his freshman comrades were trying to polish marching onto the risers to form their formation. An attractive coed caught his eye, tripping on one of the sharp stair projections. Crumpled in front of him after she had awkwardly landed on her hands and knees, he immediately reacted to her dilemma.

"Here, let me help you up. You took a real nasty spill; I hope you're not hurt."

Confused, she lay there momentarily stunned. Finally, almost frantically, she clumsily scrambled to regain her feet.

"No, I'm fine. I just banged up my knees a little. Suppose this was an act of providence?"

"How could taking a spill be an act of providence?"

"I met you. What a way to make an introduction, huh."

"Right! Are sure you aren't hurt?"

She smiled, accepting his offer of assistance.

"I knew I should have worn pants. This skirt wasn't made for a gymnastics performance."

"Your outfit may not have been suitable but I'd give you a ten for form and artistic achievement."

"Thanks for the help. You're very kind, but I feel so foolish."

Damn, she a knockout!

Her dark hair was cut just below her ear lobes, framing radiant, compelling features. She had apparently been exposed to some early season tanning sessions. As his eyes made a quick tour of the shapely treasure standing before him, he noted her figure, delightfully feminine in all aspects, reflected the trim firmness of someone dedicated to avoiding a high caloric diet. He flushed realizing his eyes had lingered too long on her noticeable qualities. Clearing his throat nervously, he slowly met her steady gaze. When their eyes finally met, her blue eyes penetrated his soul.

What do ya know? She doesn't appear to be offended by the once over I just gave her. I hope I'm not being too obvious. What I wouldn't give to see what this beauty's like when she gets all warmed up. Oh, oh, I wonder what the sly smile is all about?

"My name is Geri Hasfra, and yours?"

"Lee Grady. Yep, now you know who came to your rescue. With an introduction like this, I'm thinking, ah, maybe we should go out for coffee, after practice is over, of course. It would be nice getting to know you under different circumstances."

"You think? Since you've surveyed the merchandise, you already know what you need to know. So, which is it? Are you going to persist in saying, 'you think,' or are you going to amend your petition and admit you want to ask me out for coffee?"

"You really cut to the chase, don't you? Okay, I would really like to take you out for coffee. Better?"

"Much. I'd hate to think the fall I took was in vain."

"I doubt it was in vain. A moment ago, you called it providence. I'll take your response as a yes."

Later, in a booth near the rear of Leabold's ice cream parlor, an investigation was launched.

"Don't you just love vanilla Cokes?"

"Never tried one before. Not bad, though. So, where's home?"

"Stayton, the gateway to the Santiam Pass. What about you?"

"I was born in Bend. Ah, I've lived in Salem almost five years."

"Bend's an interesting town. One of my sorority sisters is from there. By the way, I was just initiated into the Delta Gamma."

"So I noticed, ah, your pin. Smiling sheepishly, he confessed, "I was an SAE."

"And now?"

"Ah, I ran into the first semester grade problems."

"You're not alone. We lost ten girls from our pledge class because of grades." Gently patting him on his arm, she offered, "You'll make it."

"Ah, yeah. You know, I'd like to do this again, sometime real soon."

"Share a Coke, or do you have something more exciting in mind?"

"I could be persuaded."

"Before I tie you up for the next year or so, why don't you tell me what you plan to do about your grade situation?"

"I have to attend study table at the house from six until eight every evening."

"And then?"

Shrugging, he revealed, "Usually, I just go home."

"You need to do something about your schedule. Two hours of study isn't nearly enough. We'll talk later, but maybe you should consider studying in the library with me after you get off study table."

"Sounds like a plan. Maybe we could, ah, get together and do something after we're through studying."

"Let's discuss it tomorrow, after practice. I'd like to consider what doing something might entail."

He grinned, nodding as he recalled a previous debate.

"So, are you going with anyone?"

"Not really. Too many plans for the future, I suppose. Then again, before tonight, I hadn't met you."

A few evenings later, parked in a secluded site, overlooking a panorama of bright city lights after the Glee performance, Geri pushed Lee's hand away.

"Ahem, I think it's time we took a break. You're relentless. Trying to anticipate where your hands are going to go next is impossible."

He frowned as he watched her adjust a rumpled, partially unbuttoned blouse.

"I thought you were enjoying it."

"Enjoy being mauled? You've really worked me over the past few nights."

She grinned, clearing her throat so as to restore her sense of composure.

"Maybe I'm starting to feel a little bit too good. I just need a breather to clear my head and extinguish the fire."

Grinning, he teased, "I was just getting warmed up."

"Really? I wonder why I'm not surprised. Um, suppose we should discuss some ground rules before this gets completely out of hand?"

"Ground rules?"

"Uh-huh. For the record, I'm not one of those campus hot mamma's who spreads her legs so someone can make another notch on their belt."

"I'm not, ah, I wasn't trying to force the issue."

"Oh, no? Like you weren't trying to get into my pants? Look, you're starting to cause me to lose control. You know all the right places to put your hands and all the right things to do when they're there. It doesn't hurt that you're damn good looking and very thoughtful." Grinning, she continued, "I even feel safe when I'm with you, ah, most of the time, anyway. Shhh! Let me finish. Um, I love the fact you're not full of yourself. Shit oh dear, I even love how your cheeks get red when you're embarrassed."

"But?"

"No buts about it. I could really go for you. I just want to discuss where all this is headed."

"Okay by me. Ah, where would you like for it to go?"

"How about you telling me about your intentions? Are you looking for a quick bang or something a little more permanent?"

He smiled sheepishly, ducking his head.

"I don't intend to get tied down, but I prefer a sit-down meal compared to a quick trip through the drive-in."

"I'm not looking to get tied down either, but if I'm gonna put some effort into this little affair, I'd like to think it's gonna last longer than one night." Glaring at him, she continued, "Would you have done it without protection?"

Again Lee ducked his head, his cheeks again reddening.

"Pretty risky. Since you don't plan to get tied down, what would you do if I get knocked up? Tell me you wouldn't walk away."

"I guess I wasn't thinking about, ah, I have a rubber."

"Oh, goodie. Were you planning to use it?" She shook her head with disgust. "No matter, I'm not gonna let you fuck me with a glove on, anyway. It's far too impersonal."

"Kinda reduces the options, doesn't it?"

"Uh-huh. Other than abstaining, the only one left is to practice some restraint and do it only when it's absolutely safe."

"Safe? Come on, it's never perfectly safe."

"And you don't want to get tied down? I think you're letting that thing between your legs do all the thinking." She nodded to confirm her theory. "Anyway, I'm really regular. Ah, I know when my period is going to begin, and I have a pretty good idea about the window of opportunity, ah, when it's not safe."

"You're willing to run the risk?"

"It's worked fine so far. Fuck, Lee, there you go. Your cheeks are as red as a Red Delicious Apple." Chuckling, she offered, "If you're willing, I'm going to make you my own special project."

"Hmm, sounds interesting. Care to share?"

"I'm suggesting we enter into a reciprocal trade agreement. I'll help you get squared away academically and help you improve your self-confidence. I'm not sure you really believe you can succeed. Ah, in return, you can help me avoid getting a bad case of the horns. Of course, you'd have to be willing to agree to a couple of absolute essentials."

"Such as?"

"We would both have to be free to date other people, but I don't believe in sleeping around. Catch my drift?"

"Sounds like a plan to me."

"I'm thinking we would be studying together every night. Since all work isn't my idea of fun, we'd have to relieve some tension. I think we just might get to know each other in ways you've only dreamt about."

"What's the catch?"

"We have to be completely honest with each other. For starters, ah, has there been anyone else?"

He frowned.

"Oh, come on! Everyone has a past. Who was she?"

"If you mean the girl I recently broke up with."

"Nope, she's not the one. I wanta hear about the one you don't want to talk about."

How in the hell does she know about Sandy? Is she psychic?

"I only knew her for a week."

"Long enough. Pretty special, huh?"

"Yeah. Ah, I kinda made a promise."

"Hmm. Well, I'm listening."

"I promised not to start without her."

"We're safe. Neither of us plans to turn this into a permanent relationship. Yep, I like my freedom, and it sounds like you do, too."

"I think I've heard this before."

"It's not an uncommon theme. Seriously, don't you think it's silly to say things you don't plan to honor? It would be nice if it lasted a while, though."

"How long do you have in mind?"

"How should I know? It could last forever. Until then, though, I don't want any commitments."

"Until the twelfth of never?" Shrugging, he continued, "What have I to lose? Ah, what about you? Who was the special person in your life?"

"Don't you think we've talked long enough?"

Glorious and inspiring, the warm spring initiated the birth of Lee's commitment to academics, his GPA souring by the time second semester ended.

"Geri, would you believe I got a 3.2? I really owe you."

"Hardly. You did all the work. How'd you do in English?"

"I'll read Mom Bagwell's comments, if you don't mind."

"Fire."

"She wrote, 'Mr. Grady earned an 'A' for turning an outstanding paper into a masterpiece. He has the potential to become a fine writer.'"

"I'm so proud of you. I wish you realized just how far you've come. Ah, heard anything from your fraternity?"

"I got a letter from them today. Looks like I'm eligible to go through initiation in the fall."

"No longer on probation, huh?"

"Thanks to you. Ah, what are you doing tonight?"

"What do you have in mind?"

"How about taking in a movie and going out for a bite to eat? I feel like celebrating. I'd also like to show my appreciation for all you've done for me. Besides being a good mentor, you're, ah, not bad at other things, either."

"It won't be long before our free time is defined in terms of minutes. Are you satisfied with things the way they are between us?"

"What's there not to like? Other than not seeing much of each other this summer, I doubt I could ask for more."

"Good, but just in case you're starting to think otherwise, I like our arrangement the way it is. At this stage in our lives, thinking about taking our relationship to the next level, well let's just say it holds no interest for me at all."

DOWN BUT NOT OUT

College life, a memorable transition, is a time when the exuberance and excesses of youth are gradually discarded. Attempts are made to embrace the stable maturity of adulthood while exploring new ideas and modifying behavior. A continuance of seemingly unstructured, timeless activity, never to be forgotten, it is the duration of opportunity and the quest for bonding. It is an intersection of a person's life.

"I got a job at Santiam Pack. Looks like I start Monday."
Geri frowned, obviously displeased.
"Looks I'll be working the night shift."
"I'm still gonna work at Borg's, but I'm also thinking of working in the cannery during the bean pack."
She shook her head in disgust. The scowl on her face widening.
"Our schedules stink! We'll be lucky if we ever see each other. When I'm sleeping, you'll be working and when I'm working, you'll be sleeping."
Chuckling, he reminded her, "It'll pay the tuition."
"I'll only be able to see you on Sundays, if then."
"Yeah, our day of rest."
"Humph. Think again. Damn, I'm already starting to hate summer vacation. It's been so nice being able to see you every night."
"It won't last forever."
"Maybe not, but it'll sure seem like it. Ah, I may have a solution. On your

days off at Borg's, you could come see me in Stayton."

"Yeah, a couple of hours would be better than nothing."

She grinned slyly, suggesting, "I was thinking more along the lines of spending the entire day. My folks leave for work by eight. It would give us until just before five."

"But you won't get any sleep."

"So, take me to bed. We can always sleep during break time."

At the conclusion of the summer, their deep tans had all but returned to a pale winter hue. Parked in their usual spot on a hill overlooking Salem, Geri gasped to regain her regular breathing pattern.

"Whew, I'm sure glad school's starting. Now we can be together every night."

"It's been a long summer."

She straightened up in the seat.

"Do you think I expect too much?"

"No, why?"

"You don't mind me nudging you from time to time to keep you on course?"

"I can't see anything wrong with you helping me to focus and set goals."

She grinned, playfully jabbing him in the ribs.

"And the sex isn't bad, either. I would like for you to tell me how you really feel once in a while. Would it be too much to ask to say the words?"

"Probably not. I just thought…."

"Well, you thought wrong. I'd still like to hear the words."

"I love you more than all the rolled oats in the world."

"What's this rolled oats, shit?"

"It's my way of telling you how I feel."

"Just so you know, I don't like my oats overcooked."

Geri's work wasn't done. After his initiation into the fraternity, it seemed everything he did was wrong.

"I'm starting to feel like your parole office. No sooner did you get off academic probation than you got slapped with disciplinary probation for cutting chapel and convocation. Fuck, the ink hadn't even dried and you got nailed for sprucing up the SAE homecoming sign with the logs you cut from the campus trees. Humph, but this time…."

"Driving my car on campus could cost me big time."

"What gave you the clue? Shit, Lee, the Dean won't even allow you on campus until finals. How do you expect to get the two-point the administration's requiring for reinstatement?"

"Ace my finals?"

"Really? Don't you think it's time to figure out what you want out of this place and start setting some goals instead of fuckin' around?"

"I quit the job at Borg's."

"It's a start. Look, I've invested a lot of time in you, and I love you to death, but I'd like to see all my hard work pay off, like seeing you graduate someday. Speaking of which, I've decided study languages in Paris year after next."

"Paris, France?"

"Where else? I applied for a summer institute in Vermont. If I'm accepted, my tuition at Willamette will be taken care of next year, and I'll be a shoo-in for a scholarship to the University of Paris."

"Hmm, sounds kinda final."

He shrugged, his eyes suddenly assuming a distant gaze.

"Sounds neat, though."

"It's nothing you can't do. It might help if you selected a major."

"I don't have a clue what I want to do."

"You're really good at math. What would be wrong with selecting it as your major? Selecting a major field of study doesn't prevent you from branching out. You could take some religion courses, a little sociology, and even some psychology. I think you'd make a fantastic shrink. Fuck, I'd gladly lie on the couch and let you analyze me."

"I doubt we'd spend much time with analysis unless it had something to do with studying our orgasmic frequency."

"Funny! Have you ever thought about becoming a comic? Seriously, you need to start sampling a variety of courses. You know those nightly discussions you enjoy?"

"Yeah? What about 'em?"

"The broader knowledge base you have, the better equipped you'll be to present your side of an argument."

"I learn as much from the discussions as I do in any of my classes."

"Additional exposure will make you a hell of a lot better debater."

"Come to think of it, the sociology class I'm taking came in real handy the other evening."

"How so?"

"We were discussing the pros and cons of having pre-marital sex."

"There's no doubt what side you took."

He grinned, shaking his head at her bluntness.

"Ah, one of the guys asked me to tell him what was wrong with having sex for the pure sake of having sex."

"What did you say?"

"Um, I borrowed a page out of your book when I told him most women would like to think guys are interested in more than a quick bang."

Giggling appreciably, she probed, "And, what'd he say?"

"He asked if I was doing you."

"Oops. He's kinda nosey, isn't he?"

"Just curious. He's convinced anyone going out with a girl on a regular basis is doing it."

"How perceptive of him. I guess sex is everything for him, huh?"

"Probably. I think he's still going by hand."

"Maybe I should go down and rent a milking machine for him."

"There's a problem. Cows have more than one teat. Seems like a pity to waste all the other stations."

"No problem, we'll invite some of his horny friends to join him. Then, he can be a part of a real circle jerk. Ah, if it's any consolation, my sorority sisters have been dropping a few comments, too. You should have heard them when I mentioned my bra size had increased."

"I wonder why hasn't my jock-strap size increased?"

"Good point! Why hasn't it increased? You'd think by now, you'd at least be an extra large. Gibing innocently, she probed, "Still a medium?"

"You don't know?"

"Size isn't everything. By the way, have you thought any more about looking into advanced ROTC?"

"A little, but I sure hate those one-hour drill sessions. Marching around in a hot uniform at Mac stadium and spit shining my shoes for inspection holds about as much appeal as trying to rub a wart off my big toe, but I did scheduled an appointment with the Professor of Air Science to talk about it tomorrow."

"I'd think getting a commission and attending graduate school at the University of Washington in meteorology would appeal to you."

"Yeah, but I don't want to get my hopes up. Um, I'll be a little late picking you up this evening?"

"Why?"

"There's an artist coming over to the house tonight. Art appreciation week, remember?"

The artist studied Jerry and Lee, trying to settle upon a theme for their portrait.

"Hmm, best friends, no doubt."

Jerry nodded.

"Yeah, I'm trying to convince this knuckle head how great it would be for him to apply for advance ROTC."

"Why? What's so appealing about joining some marching club?"

Jerry frowned at the artist's portrayal.

"Try getting a master's degree in meteorology in addition to becoming a commissioned officer. With a little luck, he could become a weatherman on some TV station someday."

"So what are you opting for? I suppose you want to fly."

"Yeah, as a matter of fact, I've wanted to be a pilot all my life."

The artist clapped her hands together.

"I've got it! I've been trying to think of a theme, and now I think I have the perfect motif. I'm gonna sketch your busts protruding from the body of an eagle."

"Perfect. Call it the Eagles. Now you don't have a choice, Lee. How could you even consider breaking up the Eagles?"

"Geri's been hounding me to make a career choice. Think the ROTC department would accept me?"

"Why not? It's not like everyone's lining up to join."

Late that summer, after finishing another training run in a school bus, his boss, Jim Buell offered an assessment.

"You're starting the get the hang of it, Lee. I think you'll be ready for the first day of school. What do ya say we call it a day?"

"Fine by me, ah, I have to call my girl. She just got back into town from Vermont."

"Away on vacation?"

"She's been gone all summer at language school."

Jim chuckled, clutching at his groin.

"Well, don't try to catch up too fast. She may need a little warming up. Take it from old day, these young fillies kinda like to thing giving it up is their idea. Ah, if you'd like you can use my phone."

"It would be a long distance call. Ah, she lives in Stayton."

Shrugging he inquired, "Would you like for me to dial the phone for ya?"

"That's okay. I think I can handle it."

Moments later, a familiar voice answered.

"Hello?"

"So, how was Vermont? An entire summer without being able to see you is much too long. You do recognize who this is, don't ya?"

"Of course, I haven't been gone that long. Ah, Vermont was good."

"So, how was school?"

"Hectic. How was your summer?"

"Lonely. Would you like to go out tonight?"

"I'm exhausted. I haven't had any sleep for two days."

"Hmm, is this the Geri Hasfra I know?"

"I'm just tired. I don't think I'd be much fun tonight."

"So, is tomorrow okay?"

"Come early. We can take in a movie and, ah, well, I think we have a little catching up to do, don't you?"

The next evening, parked at their favorite location in the South Salem hills, Lee frowned as he pulled away from a tentative embrace.

"You seem so distant. Anything wrong?"

"Does something have to be wrong just because I haven't spread my legs? I think we should talk about how we spent our summers."

"I'll let you start. Yours was probably more exciting."

"I hope you didn't just sit home."

"Working swing-shift in the cannery didn't provide many options."

"Come on. You must have found something to do."

"I was a good boy. What about you?"

Staring at him blankly, she mumbled, "You didn't mess around, even a little?"

"Like I said, I was a good boy. Besides, we had an agreement."

"Yeah. Um, I got the appointment to the institute in Paris. I've decided to stay an extra year and work on my master's degree."

"Which means we'll be apart for at two years, maybe more."

"Sounds like."

"What's going on, Geri?"

"A lot happened this summer. I got reacquainted with an old friend, someone I knew while I was in high school."

"Does he still live in Stayton?"

"No, he moved back east during our senior year in high school."

"Is he going to be studying in Paris, too?"

"Uh-huh."

"Hmm, do you love him?"

"Maybe."

She paused, her eyes darting to her left.

"What about the girl you never seem to want to talk about?"

"I don't know. I haven't seen her for quite a while. So, sounds like things have changed between us, or am I just misreading all the signals your emitting?"

"I'm not into doing without, and I'm not real excited about breaking in someone new. Besides, the last time I heard, we hadn't even made a commitment, right?"

"But, we did agree not to sleep around."

"True, but if you'll remember, I told you to go out with other people."

"Going out is one thing, but if you'll remember, it was you who insisted we not sleep around."

"Okay, so I blew it. I'm sorry."

"I'm sorry, too. I trusted you, but now I just don't know."

"So, I let you down."

"No, I think you let yourself down."

"Okay, so what do you want to do about it?"

"I don't know. All of this has kinda hit me by surprise."

"Look! We're still friends, aren't we?"

"I hope so."

"Don't friends fuck each other once in while?"

He grinned, shaking his head in disbelief.

"Wouldn't that make us a little more than friends?"

"Could be, why?"

"Seems to me like before you left for Vermont, we were doing it all the time."

"I never heard you complain."

He shrugged, suddenly disinterested in pursuing the subject any further.

"You're right, I suppose what we have is better than most."

"Well then, why don't we stop talking and get down to business. I think you're just about to get lucky."

What the hell, I could do worse. Besides, I'm really not ready to settle down, yet. Obviously Geri isn't either. Yeah, how bad can it be? She doesn't expect much other than getting her nightly roll in the hay. Fuck, I'd have to search high and low to find someone I'm as compatible with. Until I run into Sandy again, Geri's about all a man could hope for. Damn, I sure hope Sandy and I can get together again. Uh-huh, she did say she didn't want me to start without her, but she never said anything about relieving a little tension from time to time.

PARADE REST

In horseshoes or darts, close is good, a ringer or bull's eye a sure winner. Missing the mark, although initially disappointing, may work out for the best.

The junior year passed quickly, nondescript events blending into one another. A never varying routine foretold a ritual of getting up, driving a morning school bus route, going to class, driving the afternoon bus route, and then studying in the library with Geri. To Geri's delight, he again snubbed his nose at probation. Advanced ROTC summer camp at Fairchild Air Force Base, the first week in June, was rapidly approaching. So too was Geri's departure for Paris. Limbo! He was a hostage of a routine he had been a part of since second semester of his freshman year. Uncertainty! Two weeks after summer camp was over, he and Geri's lives would change, perhaps forever. The night before he left for camp, they sat in their favorite spot overlooking the lights of Salem.

"Want to do something really special before I leave for Paris?"

"You must mean after I get back from camp. I leave in less than a week, you know."

"Don't remind me. When you get back, we only have a week before I leave."

"So, what do you have in mind?"

"My secret. Keep your mind on what you have to do and have a successful camp. Just be prepared to drop everything when you get back. We're going to make a memory."

"It'd better be good. It'll have to last for two years, if not more."

Saturday a week after returning from ROTC summer camp, Lee and Geri strolled down the boardwalk of Seaside under a full, silvery moon.

"This has been quite a week. Did we manage to accomplish all you planned?"

"All but one. Soon as we reach the end of the boardwalk, I'd like to see if we can find a comfortable landing pad in the tall grass."

She grinned, glancing to see his reaction.

"That should just about finish off the list."

"Hmm, let me see, we wanted to wake up each morning together and eat breakfast at Pig n' Pancake."

"Yeah, I even thought it would be fun to play putt-putt golf until you whipped my butt."

"Stop complaining. You got even on the bumper cars."

She grinned, her eyes sparkling.

"The highlight of the entire week, though, was doing in just about every conceivable place in our motel room."

"Thank God we're going home tomorrow. You've worn me to a frazzle."

"Poor baby. You've really had it tough, haven't you? I hope you have a little reserve left in your tank because we have some unfinished business to take care of on the beach up ahead."

"This is going to be the last act of the play, isn't it?"

"Afraid so. Looks like there won't be any curtain calls or an opportunity for an encore performance, either."

"Another way of saying nothing lasts forever, huh?"

"You know, Lee, there have been times when I thought just maybe we might be able to turn what we have into something permanent."

"But, it really hasn't been the same since last summer."

"You're probably right. All along, I've thought I was competing with the little sweetie who broke it off with you during our freshman year."

"But?"

"Lately I've come to realize it's the gal you never seem to want to talk about. I'm up against pretty stiff odds, aren't I?"

"Sandy's special. I promised her a long time ago I wouldn't start without her."

"Humph, it seems to me you might have started without her a long time ago. We haven't exactly been practicing abstinence."

"Has everything we've shared just been about sex?"

Shrugging, she debated, "Maybe yes, and maybe no. All I know is I don't see myself living like a nun."

She jabbed him in the ribs, suddenly bolting for a sandy knoll covered with tall grass.

Shouting over her shoulder, she challenged, "See if you can read what's on my mind now, mister. Ah, it might help to hurry a little unless you expect me to do this solo."

Racing to catch up with her, she suddenly disappeared. Bewildered, he stopped.

"Come on, Geri, we're well past playing games. Where in the hell are you?"

Continuing to scan the tall grass in front of him, he called out, "Geri?"

"I'm over here. Hurry up, I'm starting to get cold."

Stumbling through the tall grass, he finally came upon her.

"My God, Geri, aren't you afraid someone will see you?"

"Now's not the time to be bashful. Hurry up and get naked."

"So this the grand finale."

"Disappointed? Hurry up, the curtain's about to go up."

The wind rustled through the tall grass where they were lying, moonbeams playing tag on his back.

"It's not the Ritz, but…."

Grunting, Lee, probed, "Is there a hidden meaning hiding somewhere?"

"I'm trying to find a reason to get excited about having sand up the crack of my ass."

"Between you and me, I like sheets better."

"The moon's pretty, though. Too bad you can't see it. I don't know if this is the time, but I'd kinda like to find out how camp went. You haven't said a word since you got back."

"Ah, no. It seems I have to slightly alter the direction I'm headed."

"What do you mean? Didn't camp go well?"

"The camp went just fine, but Colonel Wren informed me I didn't pass the eye portion of the physical."

"I don't understand. You're not gonna be a pilot."

"Makes no difference. I have to be able to pass a flight physical to get my commission. As the colonel says, rules are rules and unfortunately for me I can't get a commission without passing the eye portion of the physical."

"Bummer. So, what do you plan to do now?"

"I don't know. I don't like the idea of being thrown back into the draft pool. I may be forced to look into the Navy OCS program."

"You can't be drafted before you finish your degree, can you?"

"Afraid so."

"Well, in that case, going into the Navy is definitely the way to go." Grinning, she teased, "You'll be good at riding the waves.

"Thanks to you, I've had lots of practice."

"Want some advice?"

"Sure. Everything you've suggested has worked pretty good, so far."

"First, I want you to finish your degree."

"I can't possibly make it in four years, Geri."

"So! What's wrong with finishing in five? The important thing is finishing. If I were in your shoes, I'd forget taking French until you get all the other required stuff out of the way. During your fifth year, you could hire a tutor and take French by correspondence."

She grinned, gently pushing herself away from him. She started to get dressed.

"If I wasn't gonna be in Paris, I'd love being your tutor."

"Yeah, like we get a lot of studying done."

"I don't know. We haven't done so bad."

"This is where I say, 'It's been nice,' right?"

"We've had a good run, but it's time for both of us to move on. Um, let's head back to the hotel. If you play your cards right, we still have a little over twelve hours to test out the quality of our mattress."

"Do you want me to see you off when you fly out on Monday."

"When do you start working?"

"I'm supposed to start Monday, but I'm sure I can…."

"Makes no sense, Lee. We'll be saying our goodbyes, now."

A NEW BEARING

Changed goals necessitating an altered career path or a failed relationship all significant reasons for modifying a course to be taken in life. Neither good nor bad, the significant issue is now how the person adjusts to the alteration.

"Holy crap, Eag, you look terrible. If I didn't know better, I'd swear you just finished another bout with Geri. How long has she been gone, anyway?"

"Three weeks and for your information, nothing I ever experienced with her was as demanding as this working at two jobs. I'm absolutely beat."

"Uh-huh. What's it been? Jesus, you've been working at two jobs for, ah, twenty-one straight days, right?"

"Ten. I took a day off from the cannery a while back."

"How long has it been since you've had a date?"

"Three weeks."

"Do you have to work at the cannery tonight?"

"No, thank God. I quit last night. Too many sixteen-hour days finally caught up with me."

"Good to hear. I was afraid you were gonna work yourself into an early grave. Lately, all you've done is work. Don't you think it's time for a little fun?"

Flashing his toothy grin, Jerry suggested, "Since you're not going in to work tonight, you must be free. That means you can join Claudia and me at John's party."

"I don't think so, Jerry. I'm gonna go home and catch some sleep."

"Sleep can wait, but the party can't. Ah, there's gonna be a lot of unattached women there. In fact, Claudia's friend is gonna be there. I hear she's a real looker. Ah, I also hear she recently broke things off with the guy she's been going with. I think she's looking to get back into the swing of things. Claudia thinks you two would really hit it off."

"I don't know, Jerry. I'm bushed."

"Hog wash. Hurry up and total your charge book. I'm gonna give Claudia a call and get you all fixed up."

He winked, again flashing his toothy grin.

"It could be a hell of a party. Who knows, you might even get lucky."

"I need the sleep more than I need to meet another hot filly. Seriously Jerry, I'm wiped."

"Damn it, Eag, Claudia's friend is absolutely gorgeous. Would it help if I told you she's hot to trot."

"Excuse me. Am I interrupting an important discussion related to horse racing, or does it have more substance?"

Lee grinned, shaking his head. The unexpected intrusion penetrated the discussion.

"Saved by the bell." Wheeling around in response to the distantly familiar voice, he stared in disbelief, mumbling, "Well, I'll be damned, Diana Bodia. You're the last person I expected to see."

"Humph, usually an old friend at least says something like hello, or how are you? I guess Lee Grady operates differently."

Same old Diana, she's still sarcastic and indifferent as ever. She probably thinks she's doing me a favor by stopping by. Speaking of which, why did she stop by? Probably got bored. Then again, maybe she's slumming.

"Sorry, ah, what's it been, three years?"

"Hmm, I didn't know it had been so long."

"It's been a day or two. So, what prompted the visit?"

"You mean I don't have a right to shop in this exclusive haven for the privileged? While we're checking motives, what are you doing working here? I thought you spent most of your time at the cannery."

"Been checking up on me?" Grinning, he offered, "Working on my wardrobe." Returning to work on his charge book, he advised, "Besides, when we were seeing each other, I was working at Borg's."

Her face reddened, suddenly sensing the challenge more than she had anticipated.

"Hmm, you've changed. Usually I didn't have to bring out the heavy artillery until at least the third or fourth round."

"You taught me well. Ah, why don't you wait until I finish totaling my charge book? We can talk then."

"I suppose I could spare a few moments. I'll just browse around."

He returned to his task, glancing up periodically to acknowledge her presence.

Damn, she's good-looking. I think she's even better now than she was when we...Slow down, Lee, she's probably just shopping for her boyfriend. No problem, I need the sleep more than I need another visit to memory lane. She'd probably pick the topic, anyway.

A moment later, he approached where she was busily browsing.

"Hmm, I think men's slack would be quite becoming on you."

"Why not? Men's slacks have a certain appeal, besides, they fit better than...."

"Depends on who's wearing 'em."

"True. Um, think you could fit me with a pair of these?"

"Do you know your size?"

"Sorry, but you could take my measurements."

"Love to! Ah, you know I'll have to take an inseam measurement?"

"Don't you wish!"

She grinned, feigning a continued interest in the slacks.

"Are you busy tonight?"

"No plans other than catching up on some badly needed sleep."

"I could be persuaded to go some place for a drink and maybe a little dancing if the right person asked."

"Really! Um, how does the place on Portland Avenue sound?"

"You mean, Chuck's?"

Nodding, he confirmed, "That's the place."

"If you're asking, you could follow me home so we could go from there."

Moments later when they entered the nightclub, she quickly scanned the premises.

"Good, it isn't very crowded." Frowning, she inquired, "Unusual for a Friday night, isn't it?"

"Not really. Um, it's only nine-thirty. Give it another hour and it'll be wall to wall with people. This place usually swings."

"How about the booth over near the jukebox? While you're laying claim to it, I'm going to pay a visit to the ladies room."

"Would you like for me to order you a drink?"

"Sure, I'll have a whiskey sour. Why don't you make some selections on the jukebox while I'm gone? I feel like dancin."

"Any preferences?"

"M2, S1, and U2 are good."

"Ah ha, been here before, I see."

Grinning, she offered, "Once or twice."

Buoyant, almost celebratory when she returned, Diana announced, "Just in time. Our song's playing."

"Yeah, 'Smoke gets in your eyes" always was a favorite of ours'."

"I've missed you. I thought maybe you'd disappeared."

"You made the rules."

"Yeah. Um, you're only a year away from graduation, aren't you?"

"I still have three semesters to go. French seems to be an issue."

"I thought it was German."

She studied him carefully, methodically plotting her next move.

"Are you still going out with her?"

"No, Geri's living in Paris. How'd you know about her?"

"Small world, Lee. Ah, Geri and I are both DG's." Frowning, she continued, "I've always wished things had turned out differently."

"You have to play the hand you're dealt."

"Yeah. Things happen for a reason. I just wish…."

"It isn't like we didn't have the opportunity. So, what's been going on in your life? I'll bet you've been kept busy by the guys at the U of O."

"I haven't exactly been sitting around, but I'm not going with anyone at the moment."

"Fate. We're both unattached. Do you like it at the U of O?"

"You keep changing the subject. Afraid I'll corner you?" Chuckling, apparently enjoying the perception that she finally held the upper hand, she replied, "I like it very much. Ever wish you had gone to Oregon State?"

"All the time."

"Ah, we're both a little bit vulnerable, you know."

"What's new? We were vulnerable when we first met."

"Was it so bad?"

"I don't know if I'd want to go through it again. You have a nasty habit of not getting involved unless you're completely in control. It was interesting, though."

"How so?"

"I was just thinking about your definition of a steady relationship. Didn't you start seeing Bob, what's his name, a week after the house dance? Ah, come to think of it, didn't I hear something about you going steady with him?"

"I've lost interest in dancing. Let's finish this over a drink."

"Don't think well on your feet, huh?"

"This topic isn't suitable for discussion here."

"I have a better idea. I'd like to go when we finish our drinks."

"Why?"

"I have to be up by nine to go to work. I haven't been getting much sleep lately."

Glaring at her, now quite obviously holding the upper hand, inquired, "Tell me, Diana, what's the real reason you came to see me, bored, or is your present boyfriend out of town?"

"Maybe I wanted to see if there was any possibility we still something going between us."

"And?"

"We both know the answer, don't we?"

"Actually, anything we might have had disappeared long ago, but for what it's worth, you've always set the standard. I just hope from now on I'm able to let the ladies I meet stand on their own merit."

In the blink of an eye, Lee's senior year ended. Likewise, Diana disappeared from view, Geri stopped writing, and Sandy still continued to be missing in action. Jay transferred to Portland State and Jerry started making plans to leave for flight school, as Lee set a course to finish the one remaining obstacle to graduating. He continued to drive the school bus for the school district and enlisted in the Navy. What had been a totally distinctive experience, suddenly became a blur. Day melted into night and weeks into months. In November, Jerry wrote a long overdue letter to him. Flight school was not going well.

Hmm, looks like Jerry's coming home on leave in December to get married. Claudia finally wore him down. Old Eag better enjoy his freedom because it ends December 12th. He wants me to give Claudia a call to help her coordinate the wedding. One of the perks of being the best man, I suppose.

"Jerry suggested I call to see what I can do to help you with preparations for the wedding."

"Want to meet for lunch? We can go over everything I have to accomplish then. Besides, I have a good friend I want you to meet. Ah, it looks like the best man is going to finally get fixed up with the maid of honor."

"Am I missing something here?"

"Remember when Jerry tried to get you to go to the party at John Wold's house the summer Geri left for Paris?"

"Vaguely."

"If you had taken Jerry's advice, this could very easily be your big event."

"So, you're saying she's the real deal, huh?"

"Uh-huh."

"Why is it you're just getting around to introducing us?"

"She took a detour. Right after the party, she kinda got involved."

"Is she still involved?"

"Ah, no. She just recently ended it. Since you and Marianne are going to be helping me coordinate everything, you might as well get to know each other. Trust me, Lee. You're gonna absolutely love her."

"What makes you think so?"

"Just wait and see. Gads, there is so much to do and I have so little time."

"Relax, Claudia. Don't go getting yourself all stressed out. We'll put this puppy to bed with no sweat."

"Right, but why is it I still have doubts?"

"It couldn't be my inexperience, could it?"

"Hardly. Oh, did I tell you Jerry gets home four days before the wedding?"

"I thought he'd be here for well over a week."

"I guess he didn't he tell you he was driving home. He and that damned MG of his. I swear! I think he loves the car more than he does me. Um, you know where Nopp's is, don't you?"

"Is it the place where we're meeting for lunch?"

"Uh-huh. Marianne and I will meet your there at high noon."

"Dramatic! Do we meet out front or in the middle of the street?"

At lunch, Claudia smiled, noting the electrical impulses darting between the two principals of the wedding party.

"I think you two know what has to be done. I have to run, so, I think I'll leave you to sort things out."

"I have to go, too. Lee, if you'd like, you could call me tomorrow so we can get started. Ah, my phone number's in the book. It's the only Sields listed. Claudia, would you mind waiting a minute? I have to go to the ladies room, but I'd like to talk with you about something."

Claudia grinned, giving Lee a wink.

"Don't be long, I have a million things to do."

Giving the thumbs up signal as Marianne disappeared into the restroom, Claudia boasted, "Didn't I tell you?"

"You didn't lie. She's really something, but, ah, you know I'm not really looking?"

"I hope you're not still mooning over Sandy, what's her name."

"I don't know I'd call it mooning, but I still think about her from time to time."

"Still wouldn't hurt to keep an open mind, would it?"

"I suppose not. Yeah, taking Marianne out wouldn't be so bad."

"You'll be glad you did. Give her a call. I think she's interested."

The next day when he called, Marianne's voice had a receptive but mysterious ring to it.

"Oh, my goodness. What a surprise, Lee, you called."

"Uh-huh. As I remember, you suggested I give you a call. If that's a surprise, then...."

"How would you like to get together this evening to get organized?"

"Sure. My list is relatively short, but I could give you a hand with yours."

"If getting together is too much trouble, ah, we could discuss it over the phone, Lee."

"I won't be needing any supervision, but some company would be nice. Isn't it called mixing business with pleasure?"

"We might even get to know each other a little better."

"What an intriguing idea. Ah, this evening would be fine."

"How does seven sound?"

"Musical. If we make it a quarter past eight, we'll have the lyrics of a song to guide us."

"You could be dangerous. Hmm, Lee Grady, where have I heard your name before? I've been trying to place it since I met you yesterday."

"The only people I know of that are attending the U of O are Mel Marsters and Diana Bodia."

"Sure enough. Diana and I are sorority sisters. You wouldn't also know Sandy James, would you?"

"Yeah, but how do you know Sandy? She should have graduated before…."

"I was on the five year plan. She was a senior my freshman year. I'm a year older than you and Claudia. Um, you must be the guy she talked about all the time. You made quite an impact on her, Lee. She talked about you all the time."

"How is she? I haven't seen her since…."

"She got back from Europe a short while ago. She has lived over there since she graduated."

"Is she still planning to go to law school?"

"I really don't know. Um, back to the issue at hand. I'm really envious of the small number of things you have to do. Of course you realize most of it can wait until Jerry gets home."

"I know. All I have to do right away is set up the bachelor party."

"Where are you thinking of holding it?"

"The Marion Hotel."

"Good choice. Need any help?"

"Sure, would you like to go with me to make the reservations tomorrow?"

"I'm between jobs, so anytime would work. Good thing, too. This wedding would have been hell to plan if I was working."

"Oops, two days in a row. Sure we're not pushing things too fast?"

"I'm not in any hurry to get involved, but I think we can handle seeing each other. Just because we decide to go out, it doesn't mean...."

"Right. Ah, I'm not looking to get involved, either."

"Relax. I think it's gonna be fun just getting to know each other. We might even enjoy the ride."

The bachelor party was a hit, booze flowing like a fountain and all of Jerry's closest friends in attendance. A surprise visit by an exotic dancer was the hit of the evening until a showstopper performed the final act. Inebriated to the max, Lee brought down the house attempting to seduce a cement mermaid at the bottom of the swimming pool.

The next day, scurrying around to take care of last minute details, Jerry massaged his temples while he sat slouched in a booth at the Windmill restaurant.

"Tough night. So, Eag, do you think I'm making a mistake?"

"Isn't it a little late to have doubts?"

"I want to marry her, but I'm just wondering if I'm ready. I was really starting to enjoy the freedom of being a single Air Force officer."

"You're a very lucky man, Jerry."

"I know. It's just.... Oh, what the hell. Let's go over the list again. The reservations have been made for us at the Benson Hotel, correct?"

"Yep, and your MG is packed and ready to go."

"Do you have the ring?"

"I put it in the pocket of my tux."

"What about the check for the minister?"

"You need to relax. The check's in my wallet."

"What about my gift for Claudia?"

"In my car. There's nothing else on the list to be taken care of except for you and Claudia to do your thing. Jesus, Eag, you're making me nervous."

In the reception line, following the ceremony, the signals darting back and forth between Lee and Marianne were unmistakable.

During a lull, Jerry teased, "I hear you and Marianne are starting to heat it up. Anything I should know?"

Nodding uncomfortably, Lee confessed, "Marianne's an intriguing woman."

An hour after sending the new couple off, in the hills south of Salem, the brilliant lights of the city illuminated their bedroom.

Bending forward to kiss Marianne's cheek, he whispered, "Time to go."

"Already? Suppose Jerry and Claudia made it to the hotel yet?"

"Unless they got lost on the freeway."

"We did a fabulous job of getting everything ready for them. I can't think of a single detail we overlooked."

"We even turned down the bed and tried it out."

Her eyes twinkling, she purred, "If our night together's an indication, theirs should be outstanding."

Marianne playfully jabbed him in the ribs, glancing in the back seat.

"Hmm, were you planning on spending the night here?"

"No, why?"

"What's the suitcase doing in the backseat, then?"

Lee's face paled. Emphatically, he slammed his fist into the seat.

"Damn, I forgot to put it in Jerry's MG."

"Of course you're kidding."

"I wish I was. So, what did you do with Claudia's bag?"

"I gave it to her dad. He was supposed to put it in Jerry's car."

"How? I had the keys."

"Are you serious?"

"Deadly. We better find out what he did with the bag, and then, somehow, we have to get the suitcases to them."

COMMENCEMENT

The ceremony commemorating the end of an appointment with the world of academia is cause to take note. The observance, not always conspicuous, nonetheless celebratory, gives cause for those who have supported the effort to admire the document denoting completion of the years of toil. Resting in its sheath, it signifies the bearer has reached a state of preparedness to seek advanced grooming in a more specialized field. Unlike high school commencement, there won't be teary farewells. College comrades merely mutter hasty goodbyes, a half-hearted encouragement for some future reunion. Occupied with thoughts about the world of economic adventure, social service, or perhaps a military commitment, they take little time to reflect upon the toll exacted by the academic travails. No time to recall the freelance frolicking during idle moments where they escaped the pressurized world of scholarly pursuits. All things come to an end. The future looms on the horizon, challenging and daunting.

After the graduation ceremony, Ron Tielson, a friend Lee had worked with at Meier and Franks, joined him and his family at their home for a celebration dinner. From the onset, Tura was in rare form, too much rum causing her to nearly lose cognizance of her actions even before serving dinner. Euphorically wasted, her moment of pride, like the tide, ebbed and flowed as she gradually eased into a subliminal commemoration of the event.

"Today is a dream come true, and I'm damn sure going to enjoy it."

The manner she served dinner said it all. Mere micro-moment after setting out a marvelous entrée of turkey and all the fixings, a grin plastered on her face, she rose and headed for the kitchen.

"Who wants dessert? I'm serving angel food cake."

While everyone suppressed an urge to snicker, Addison, in his typical stoic form showed no particular concern or disapproval, merely shaking his head. Hastily, he began attacking the food on his plate.

"You'd better hurry up and eat before she decides to clear the table."

Teetering unstably, a curious grin spread across Tura's face.

"What's the matter with all of you? You're not all gonna be party poopers, are ya?"

She set the cake on the table and turned to give Ron a big hug.

"Ron, I thought you'd like a little sweetness with your turkey."

Chuckling with unrestrained humor, he retorted, "Tura, I want to thank you for inviting me to this wonderful dinner. You have a lot to be proud of."

"It was a long time coming, Ron. I hope you get busy and get your degree."

"I'm thinking of enrolling at OCE this fall." Turning towards Lee, he probed, "Anyone you want me to keep tabs on while you're gone, Eag?"

"Nope. I finally put Diana to rest and Geri is getting married, so I guess there's nobody else to worry about."

"Hear anything from Marianne?"

"Yeah, I'm stopping in Detroit to see her on the way back to OCS."

"What about Sandy, what's her name?"

Lee frowned, shrugging nonchalantly.

"I don't know. I haven't heard from her. She's probably found somebody else."

Later, as Lee and Ron said their goodbyes, Ron gave his friend a hug.

"Keep in touch, will ya?"

"The mail travels both ways, Ron."

"Thanks again for inviting me. Your mother is a panic, and I absolutely love her to death. It's amazing how Addison manages to sit there, smiling, hardly ever saying a word."

"Yeah, Mom and Dad are quite a pair. Ah, do ya think I'm being overly sensitive if I confess to being a little nervous about the OCS gig that I've signed on for?"

"No. I guess if there is some good news about what you're gonna be doing, it's the opportunity you'll have to do a lot of traveling."

"Yeah, join the Navy and see the world, huh?"

AN UNEXPECTED ENCOUNTER

A drop-in visit, often a surprise, is also often unannounced. Sometimes awkward, it can also be very pleasant. Sometimes preceded by fantasy, the inner voice wonders, "What if." Sometimes it even says, "I wish."

Two days before leaving to report for duty at the Naval OCS training facility in Newport, Rhode Island, Lee made a visit the bowels of Salem's commercial center to pick up a few things for the trip. Standing next to his car parked in front of the Marion County Courthouse, he was surprised by the familiar sound of Colonel Wren's voice.
"Lee, Lee Grady, wait a minute, I need to talk to you."
Lee smiled as he watched the colonel approach from the other end of the courthouse.
"Colonel Wren, it's good to see you. It's been too long."
"You're right. I haven't seen you since our meeting after summer camp nearly two years ago. It's fortunate I ran into you."
"How so?"
"The Air Force changed its policy. You can be reinstated into the ROTC program and get your commission."
"No kidding. You don't know what I would have given to hear those words a year ago."
"And now?"
"I enlisted in the Navy. I leave for OCS in two days."

The colonel frowned, disappointment etched on his face.

"The nineteenth of June?"

"Doesn't give me much time, does it?"

"It's not too late, you know. I could pull some strings and have your enlistment in the Navy voided. I still have a slot for you in ROTC."

Lee eyed the colonel carefully, thoughtfully, countering, "I appreciate the offer, but I've made a commitment. You always did teach us to honor our obligations."

A distantly familiar voice interrupted, "Did you honor yours?"

Lee wheeled around towards the sound, his eyes coming to rest on the enticing, shapely form he had dreamt about so often.

"Sandy?"

"Sorry for interrupting, but there's something to be said about honoring a commitment. It always gets my attention."

"Sandy James. My goodness, you look great. Ah, Sandy, allow me to introduce you to the Professor of Air Science at Willamette University, ah, Colonel Wren."

"Lee, I have to be going. I wish you the best. I'm sure you'll make a fine Naval officer. Ah, nice to meet you Miss James."

My God, she's more beautiful than ever. How many times have I wished for this moment? Now, I, ah, I don't know exactly what to do. I wonder what she's doing now. Suppose she's still unattached?

"I don't believe it. It's really you."

"In the flesh. Yes, it's me."

"How are you? My God, how long has it been?"

Giggling excitedly, she ebulliently exploded, "I'm fine. Ah, how does about five years, three and one-half months sound? God, you've changed, Lee. You've become a man. You always were handsome, but now.... You don't know how often I've dreamt about a moment like this."

"Me, too. I can't count the number of times I've thought about you, and, ah, Sandy, you're as beautiful as ever, even more so. What are you doing in Salem? I thought you'd be living any place but here."

"I'm an assistant DA for Marion County. I've been on the job for about a month now."

"Wow! I always knew you would make it big. How about coffee, or do you have the time?"

"Make it a drink, and you're on."

Grinning as she stole a quick glance at her watch, she suggested, "I don't have a thing on my calendar for the rest of the afternoon. If you'll give me a minute to check in with my office, I'll meet you in the lounge of the Senator Hotel. Um, on second thought, let's make it ten minutes. I don't want to be late. Smiling, she teasingly scolded, "Ah, just don't start without me?"

Ten minutes later as Sandy made her way to his table, an alert hostess followed closely behind.

Waiting patiently until she was seated, the waitress inquired, "What'll it be? Just drinks or would you like a menu?"

Surveying the table, Sandy noted, "I see you're drinking beer. My habits have changed a little since I saw you last. I think I'll have a whiskey sour, tall glass with lots of ice."

She hesitated, waiting for the waitress to leave.

"I assume you got your degree. Is what I overheard, true? You've going into the Navy?"

"Yeah, I'm leaving in two days for Navy Officer Candidate's School."

"It's a real coincidence I ran into you. I was going to call you."

"You look fabulous, Sandy. God, I'm glad…."

"Me, too. Funny how things work out."

"What do you mean?"

"Just when I was thinking of looking you up, you're getting ready to leave. Ah, you don't know how many times I've wished I had stayed."

"It was for the best, Sandy. We weren't ready."

"And now?"

"It seems like it's always the wrong time, right place or…. Where did you get your law degree?"

"Willamette. I just passed the bar exam."

"You're kidding! I, ah…."

"I know. I've been following your activities. What do you hear from Geri?" Grinning, she pressed on, "And how's Marianne?"

His face reddened.

"Marianne mentioned you were sorority sisters, and Geri's getting married. What about you, are you…?"

"Single? I've been too busy to get myself tied down, besides, I, ah…."

"I heard you lived in Europe after you graduated."

"I took a little time off after graduating, three years to be exact."

"Really. Where did you live?"

"All over. I guess I was looking for some answers."

"Did you find them?"

"I think so."

"And now you're a lawyer. Ah, if you knew I was at Willamette, why didn't you give me a call?"

"I considered it. You don't know how close I came to calling you, but, ah, every time I was about to make the call, ah…. I've really missed you. God, you don't know how much I've missed you. Is it serious with Marianne or did you decide to not get started without me?"

"We're just friends. Is there anything you don't know about me?"

She smiled, toying with her drink.

"How's your friend Mel?"

"He's great! He has been going to school in Eugene. I think he's thinking of finishing up at OCE, though. How's your sister?"

"Married. She's the proud mother of two beautiful children. Strange how everything works out, isn't it?"

He nodded as he studied his drink.

"It is, isn't it? Thinking about what might have been sort of boggles my mind."

"Please don't. I felt so badly about what I did to you. I doubt there was much more than a day at a time the past few years I didn't wonder if I'd made the right decision."

"You did. As I said, we weren't ready. At least, I wasn't."

"So, are you planning to make the Navy a career?"

"I plan to put in the least amount of time possible. You planning to become some big wheel in the criminal justice system?"

"Sure, why not? Actually, I'm not certain about my career path. For the first time in my life I really don't know what I want to be doing in two or three years. I'd like to become a federal prosecutor, but getting there is going to be real tough."

"You'll get there, Sandy."

He smiled as he ran his finger around the rim of his glass.

"You could always wait for me to get out of the service."

"And then?"

"Hell, I don't know. It was probably a stupid suggestion."

"Was it as stupid as what I am about to suggest?"

"Try me."

"Order another round of drinks while I go to the main desk to get a room. After we finish our drinks we can catch up on some lost time."

"Is that what you really want? I'm not sure these five-year reunions are good enough. I've always hoped there'd be more."

"For now, it doesn't look like we have a choice."

"I suppose you're right. Yeah, go on and get the room."

A few minutes later the cocktail waitress returned with another beer.

"The lady already paid for your drink. She told me to tell you she decided to go home and wait. She also told me to remind you not to get started without her. There was also something about not wanting to ruin a beautiful memory."

She smiled, stepping back from the table.

"For what it worth, honey, if I were you, I'd take her offer real serious. She seems like the real deal."

"She is. Yep, she's real special."

"Well, there you go." Winking, she continued, "Just remember, young man, if anything comes of all this, you owe it all to me. Yep, I rarely miss a call on situations like this."

A RENDEZVOUS

Can you ever go back? Is a return engagement ever the same? Time and new experiences change us all.

Lee hugged his mom as the call to board echoed around the boarding area at PDX.

"I'll write as often as I can, Mom. Ah, please don't cry. I'll be home before you know it."

Turning towards his dad, he extended his hand.

"Sixteen weeks isn't too long, kid. Take care of yourself."

"I will."

Winking at Greg, he challenged, "Give 'em hell at North. I'll be anxious to hear how you do in wrestling and baseball."

Aboard United Airlines flight 562, he settled into a seat by the window awaiting takeoff. Moments later the airplane was taxiing down the runway. A first flight, he looked out the window, anxiously awaiting liftoff. Seconds later the plane was airborne.

Let's see. After I get to Chicago, I'll transfer planes. According to my ticket, I should land in Ypsilanti about six-thirty. Marianne said her place in Ypsilanti was only a few minutes from the airport, so I should be able to get a little sleep before we take in the Yankee-Tiger game. I wonder how far Ypsilanti is from Detroit?

A few minutes later the Captain's voice came over the intercom.

"On behalf of the crew, I'd like to welcome you to United Flight 562 and thank you for allowing us to be your carrier. When we reach altitude, we'll be cruising at 32,000 feet. Looks like we've picked up a tail wind, so we should arrive in Denver a little ahead of schedule. The weather in Denver is clear and seventy-degrees. When the smoking and the seatbelt lights turn off, you'll be free to move about the cabin. Enjoy your flight, and again, thank you for flying United."

Wow, this sucker doesn't waste any time. One second you're on the ground, and the next you're in the air. I wonder how much longer before we reach altitude? Good to hear the weather's nice in Denver, but I wonder how it is like over the Rocky Mountains? I hear it can get a little rough. It was really good to see Sandy, again. Nothing is certain, but at least it sounds like there's still hope. After seeing her again, my feelings for her were confirmed. I'm still in love with her.

He cleared his ears and settled back in the seat. Marianne had wrangled two days off from her busy flight schedule, convincing him to leave three days early so they could spend some time together before he reported for active duty. She had written that they would get in lots of sight seeing and catch a Yankee-Tiger baseball game. Her roommates were going to be out of town so they would have the apartment all to themselves.

He closed eyes, recalling the last time they were together before she left to attend stewardess school. A newly married friend had invited them to a party at his West Salem apartment. For unexplained reasons, they argued over everything. Midst one of their many disagreements, Marianne announced that she was going to stewardess school for American Airlines. She was going to be leaving in two weeks.

"Nice of you to let me know."

"Don't get huffy, I just received the acceptance letter today."

"How long have you known this was a possibility?"

"I applied before we met. I know I should have told you, but it completely slipped my mind. It's not like I haven't been busy, you know?"

"Yeah, well I applied for Officer's Candidate School before we met, too. It didn't keep me from telling you."

"Oh, boy. I can see tonight's gonna be something else again. I don't know why we came to this party in the first place. I started my period, and I'm in a real bitchy mood."

"Tell me something I don't already know."

"Fuck you! If you're not happy, why don't you hook up with the little sweetie over in the corner? Every time I look over there, she appears to be giving you the eye."

"You mean the little blonde with the big jugs?"

"Go for it if you think they're so special."

"This evening's going nowhere. Let's get out of here."

"I plan to stay for a while. I'll find a ride home if you want to go."

"Well, enjoy the rest of the evening."

"Yeah, you too. Don't trip over the cute little blonde on your way out. I hear she's easy."

"Hmm, how novel."

"Oh, so now you evaluate the quality of a sexual experience based upon how difficult it is to come by. I thought we had a pretty good thing going, but maybe I was wrong."

The disagreement exploded again the next day when he called.

"I'm sorry about last night. You really caught me off guard with the news about going to stewardess school."

"I think we should stop seeing each other for a while. I don't want to go to stewardess school worrying about a relationship back home."

"What do you mean by a while?"

"I don't know, but I want to take a little break until I finish school."

"Probably not a bad idea, I still have some classes at Willamette to complete. Um, do you want to write?"

"Do you?"

"I'll write if you do."

"Friends?"

By the time Marianne pulled into the driveway of her apartment, Lee was fast asleep.

"We're home, Lee. Wake up. Come on, honey, I'm gonna need some help with your bags. I can get the small one, but the big one looks like it weighs a ton. What's in it, anyway?"

He grinned, lifting his tired body out of the front seat.

"Comic books."

Outside the car, still struggling to wake up, he stretched and then rubbed the sleep out of his eyes.

Yawning, he mumbled, "I'm absolutely wiped. All I've done the past few hours is fly and change planes. So, we're home already, are we? I hope you're gonna let me catch a couple hours of sleep before we go to the game."

"I figured you'd need a couple of hours. Ah, we'll head into Detroit to catch the game about noon. Looks like you'll get a little bonus."

She smiled as she watched him head for the front door of her apartment.

"It's unlocked. Just leave your bag in the living room. The bedroom is down the hall, first door on the right."

The mask of sleep slowly retreated. His eyes fluttered in response to a tender announcement proclaiming the unconscious trance enveloping him had endured long enough. The warm, soft figure beside him seductively purred.

"I got lonely waiting for you to wake up."

Instinctively he reached out to encourage her movement towards him, lightly touching her right shoulder to announce a readiness to receive her presence. Bare, silky flesh immediately stimulated him, blood surging to his temples. A quick glance confirmed her state of nudity. Beautifully sculpted breasts, alluring jutting mounds with dark brown swollen caps, loomed through his blurred vision. Partially hidden under the sheet, they rose and fell rhythmically in a synchronized harmony with her steady breathing pattern.

"I'll be back in a moment. I've got to go to the bathroom."

"Hurry back. I'm already missing you."

After giving his toothbrush a workout, he guided his body into bed beside her.

"Miss me?"

"I thought you'd never return."

She greeted him eagerly, responding without resistance to his bold, caged passion. Deep breathy gasps escaped her parted lips enunciating her increased emotional readiness as he artfully began to fondle her breasts.

"I've missed this."

"Me, too."

Slowly he guided his hand between her legs, gently exploring her damp, responsive harbor.

She stiffened, suddenly, unexpectedly pushing his hand away.

"Not yet. I'm sorry. I'm just not ready, yet. I need some time."

"Humph," he moaned as a disappointed rush of air escaped his mouth. "I guess you want to get reacquainted first."

"I'm sorry. I didn't mean to lead you on. Um, let's get ready to go to the game. Yeah, maybe we can pick up where we left off later this evening."

The visit to Tiger Stadium, a fulfillment of a lifetime dream, had been a fantasy since hearing his grandfather's tales of the Yankee's Murderous Row and the great Bambino. New heroes had replaced them but not the mystique of the immortalized pin stripers. It hadn't been easy to be an ardent Yankee fan, filing a minority report while being forced to root silently. Seated midst all the Tiger fans in Tiger Stadium, he again recognized his role.

"It might be a good if I kept my mouth shut, huh."

Marianne grinned, nodding.

"Unless you're a Tiger fan."

The game moved into the top of the seventh inning with the Tigers leading 7-4, the bases loaded. The sky opened, dropping a torrent of rain onto the playing field. Moments later the umpires halted the game and signaled for the ground crew to cover the field with massive tarps.

"Damn! My first major league game and it gets rained out."

"No way. The umpires are just delaying the game until the storm blows over. For your information, a storm like this is quite common. Ah, it usually only lasts for a few minutes, an hour tops."

Thirty minutes later as they sat discussing the Tiger's season with some of the surrounding loyalists, Marianne elbowed him.

"See, it's letting up already. The umpires just gave the ground's keepers the green light to get the field ready. The game should resume in another thirty minutes."

"Good! I'd hate to miss an opportunity to see the Yankees take care of business because of a little rain."

"Shhh, Not so loud. Have you forgotten whose stadium you're in?"

When the game resumed, Yogi Berra stepped to the plate to face the Tiger pitcher, Denny McLaine. Oblivious to being in hostile territory, Lee could contain himself no longer.

At the top of his lungs, he shouted, "Come on, Yogi. Knock one out of here." Turning excitedly towards Marianne, he explained, "Look at the way he waggles his bat. He's ready to deliver."

He leaned forward with anticipation, patting her knee.

"Yogi's a favorite of mine. Come on, Yogi, park one."

"Strike one," the portly umpire bellowed.

Yogi reached down and grabbed a handful of dirt, rubbed his hands together, and then again assumed a position in the batter's box. As he readied for McLaine to serve up the next offering, he took two deliberate practice swings, symbolically spitting into the dirt.

The umpire's second pronouncement echoed through the stadium, "Strike two!"

Lee groaned, "I can't look!"

"Don't worry, honey, Yogi wouldn't dare let you down."

With the count 2-0 on the Yankee catcher, McLaine reared back and unloaded his next offering. The ball rocketed towards the plate, noticeably high and outside.

Without hesitation, Yogi reached out with an awkward sweeping motion of his bat, and simultaneously an indescribably beautiful sound echoed throughout the stadium. The Tiger pitcher's ninety-mile-an-hour fast ball ricocheted off the squat slugger's smoothly sculpted weapon, screaming towards the center field wall. Rising higher and higher, it disappeared into the bleachers.

Gone!

Standing in awe, Lee watched Yogi circle the bases.

"Can you believe it? It's out of here. Shit, this is like a fairytale."

Marianne smiled, shaking her head.

"One huge difference, honey. To the Tiger fans, this is real. They've experienced this scene too often." Squeezed his arm affectionately, she probed, "After the game is over, what do you think about going to a favorite restaurant of mine?"

"Sounds good. I'm so hungry I could eat a horse."

"I doubt you'll find it on the menu where we're going. They serve Chinese."

"I like your choice of restaurants. Chinese is a favorite of mine."

"Good. Afterwards, I think we'll head back to my apartment. You may still need to catch up on some of the sleep you lost." Grinning, she teased, "Course, if you're not too tired, we might even find a little time to mess around a little."

"You sure? This morning you seemed a little reluctant."

"So, ah, now I'm in the mood. What about you?"

"Always, ah, especially with you. What's on the agenda for tomorrow?"

"What do you think about me serving as your tour guide and showing you the sights of the Motor City?"

"Sounds good. I'm sure there's more to this place than what I've seen so far."

After dinner and a short drive back to Ypsilanti, they pulled to a stop in front of her apartment. Marianne frowned as she turned off the ignition, gazing intently on the late model sedan she had parked beside.

"Hmm, that's strange. If I didn't know better, I'd say that's Janet's car. I wonder what's up? She's supposed to be on a two-day flight to New York. Um, this may change our plans a little. Oh, well." Chuckling, she challenged, "Ah, last one to the door has to buy breakfast tomorrow."

Bursting into the apartment, slamming the door behind them, they playfully grappled with each other.

"Ahem, am I interrupting something?"

Giggling as Lee continued to tickle her, Marianne, gasped, "Janet, ah, I thought you were scheduled for a two-day flight."

The comely blonde chuckled, amused at the playful display.

"Obviously. You'll have to excuse me, but I was just putting the finishing touches on a snack."

Still entwined in Lee's arms, Marianne smiled sheepishly.

"What happened to the layover flight to New York?"

Stuffing the last bite of a sandwich into her mouth, Janet paused for a moment to finish chewing. Finally done, she waved her arms frantically as though clearing a path for a response.

"The trip got scrubbed, engine problems. They've rescheduled the flight for tomorrow. Wouldn't ya know it? I have to be at the airport by six in the morning. Guess the plans Roy and I made for the weekend are in the toilet."

"I'm sorry about that. Um, Janet, I want to introduce you to Lee Grady. He's the friend from Oregon I've been telling you about."

Lee nodded at Janet, a sly grin spreading across his face.

"Aren't I lucky? Marianne has so many she has to label them. You know, Oregon, California, Michigan…. Anyway, it's nice to meet you, Janet."

"The pleasure's all mine. Until I met my roomie, I never ever knew anyone from Oregon. Now I have the pleasure of knowing two people from the Beaver State."

"Where you from, Janet?"

"I'm from the great state of Texas. You know where everything is bigger and better than anyplace else in the entire world."

"Yeah, yeah. In case you didn't know it, Lee, Texans have the market cornered on bullshit. Let me tell you what's really big." She winked at Lee, continuing, "Those damned Yankees are really big."

"Ladies, I'm beat."

Lee stretched as he yawned audibly.

"This cowboy is going to call it a night and hit the hay. Ah, which room do you want me in?"

Janet chuckled, returning to her sandwich.

"Come on, Lee. You know better than that. Take the one with Marianne in it."

Marianne's face reddened, overruling her roommate, "You can have the room you slept in this morning. I'll move my stuff to Trudy's room."

Lee smiled, nodding at Marianne's roommate.

"It was a pleasure meeting you, Janet. I hope you have a good flight tomorrow."

He smiled, turning toward Marianne.

"See you in the morning."

"Don't dally while you're heading down the hall or you might get run over. This little stew is headed for the rack and a few winks before I have to hit the air lanes again."

Lee jerked to a state of alertness as the engine of Janet's car roared to life. He blinked with confusion, trying to determine where he was. Slowly the soft presence of Marianne's body captured his attention.

"I thought…."

"Well, good morning to you, too. Ah, I like sleeping in my own bed."

He kissed her lovingly on the forehead, gently encouraging her to come closer.

"So, when do you want to leave."

She smiled coyly, suggesting, "Oh, I think we have a little time before we have to leave for the big city, don't you?"

He leaned forward, greeting her soft lips. After a moment he broke from their embrace.

"Want to take a shower first?"

Just before ten o'clock, they left her apartment.

"We're gonna be driving to the same place where we parked the car yesterday, so if you don't mind, I'll let you do the driving."

He grinned, accepting the keys from her.

"As long as you'll act as the navigator and give me directions far enough in advance. This place is still kinda foreign to me."

"No problem. I've picked some neat places I'd like to show you, but I think we'll start with my personal favorites. I think you'll like the places I've picked for us to visit."

"Any chance we're gonna tour an automobile factory?"

"It's on the list. Take the next exit. When you're on the freeway, you're gonna take the fourth exit. I think when you turn off the freeway, you recognize where we are."

"After we park the car, are we going to be walking or using the bus?"

"We can walk to the first place I'd like to show you. We'll use the bus afterwards." Nodding her approval, she revealed, "This is a real treat. I like it when someone else is doing the driving. Besides, with you driving, it will give me a chance to tell you some stuff I think you deserve to know."

"Hmm, sounds ominous."

"Somewhat. Ah, I met someone."

Pausing to let the news settle, she toyed with an air freshener hanging from the rearview mirror.

"Ah, about now I suppose you're wondering why I suggested you visit me before reporting for active duty?"

"Oh, I don't know. Besides getting together with a good friend, maybe you wanted to see if we still have anything going. How am I doing so far?"

"Not bad." Frowning she probed, "You don't seem to be upset about what I just told you."

"Am I supposed to be?"

"A small amount of concern might be good."

"Whether you've found someone else or not, it doesn't stop us from being friends, or does it?"

"Sure, I mean, no. Ah, I hope we'll always be friends, but right now, I'm just trying to determine what kind of friends we'll be."

"Meaning?"

"Well, some friends hang out with each other. Others hug and kiss. Really good friends fuck each other."

"If this morning was any indication...."

She blushed, ducking her head.

"Old habits are hard to break. Ah I'm still trying to decide what to do about you."

"I don't mind just being your friend, Marianne. I don't think I'm quite ready for anything deeper, just yet. Ah, why don't you tell me what a person does for an entire day in Detroit?"

"They spend the day visiting museums and art galleries, walking up and down the streets of the busy commercial center, and poking their noses into far too many stores for anybody's good."

"Well, you're the guide. Just be gentle. Too many shops could leave me bruised."

When an exhaustive search of Detroit's nooks and crannies concluded, they welcomed the return to the soft seats of her car.

"I'm beat. I'm sure glad you decided to call it a day. As far as shopping and snooping about goes, you're a Goddamn athlete, Marianne. You must have showed me every hideaway Detroit has to offer."

"I could have shown you more, but you did ask me to be gentle."

Another surprise greeting awaited them at the apartment. Trudy, her other roommate had also experienced the unpredictable nature of the flight industry. After brief introductions, Trudy joined the tired tourists in the kitchen for some refreshments, her humorous account of a previous overnight flight to Seattle thoroughly entertaining them. Suddenly Trudy looked at her watch.

"Heavens to Betsy, I didn't realize it was so late. Can you believe I have to be at the airport by five-thirty? Well, Trudy's gone."

After she had disappeared down the hall, Marianne's eyes briefly met his as they sat in silence. Visually they discussed the prospects lying ahead of them, sparks dancing about the room as he eyed the path to her bedroom.

"Trudy will be asleep before long. Unless you're tired...."

"Not this cowboy. For you, I can always be ready."

Strange how the mind works. One minute, reality exists in the present, the next it drifts lazily along in some time warp. Sometimes a sleep-induced coma can expedite the venture.

Roderick crept towards on the house resting on the hill. Located across an alley behind his home, it was rumored to be a sanctuary for an evil witch. The mysterious, foreboding shack was now so close, he could almost sense her presence within.

"Have you ever seen her?"

"Are you kidding? I don't think anyone has seen her."

Lee nudged his friend, signaling for him to continue moving towards the old lady's house.

"Stay low while we're sneaking up on the porch so she can't see us. Ah, are you gonna knock on the door?"

"Sure, but you do the talking. I wouldn't know what to say. I don't know if I like the idea of holding her up or not. The idea makes me shiver."

"Why? What's she gonna do, anyway? She's gonna be so surprised she'll probably fork over all her loot just to get rid of us."

"Maybe so, but I keep getting the creepiest feeling. Are you sure she won't put some spell on us?"

"What's the matter, lose your nerve?"

"Nah, I just want to be sure I know how this is all going down."

"Well, knock on the stupid door and stop messing around."

Roderick doubled up his fist and started banging on the creaky screen door enshrouding the front door of the run down shanty. The poorly maintained dump serving as Miss Penrose's home was in desperate need of a new paint job, other necessary repairs too numerous to mention. From within, they could hear heavy footsteps approaching the door.

"Oh God, here she comes! Quick, hide over there on the other side of the screen door. I'll take this side."

The door opened.

Who's there? Hmm, must have been the wind, but I'd better check and see for sure. It could be some of those unruly brats playing tricks on me again.

The mysterious Miss Penrose opened the screen and peered out.

"What the…? What do you young rascals have up your sleeves?"

"Stick 'em up!"

Lee raised Roderick's BB gun to waist level and pointed the barrel directly at her heart. Somewhat startled, she sneered in an attempt to suppress a smile.

"What do we have here? Well, come on. I don't have all day. What do you want, my money or what?"

Roderick closed in from the other side.

"We'll take all your candy!"

The old lady laughed.

"Why don't you boys come in and we'll talk about this matter over some milk and cookies."

Roderick frowned, looking hopelessly at Lee.

"Is this how it's supposed to work?"

Lee shrugged as he lowered his gun.

"I guess it wouldn't hurt to have a cookie or two while we case the joint."

The old lady showed them to her kitchen table and brought a large platter of chocolate cookies. She set the platter on the table and nodded.

"You desperados eat up while I fetch the milk. What's your pleasure? I mean, do you want chocolate milk or the regular kind?"

Lee struggled to clear his mouth so he could answer.

"I like chocolate."

"Me, too."

Mrs. Penrose nodded as she watched the cookies disappear.

"Aren't you the son of the man who owns the market down on Newport, below my house?"

"You won't tell my dad will ya?"

"I suppose I could be persuaded to forget the matter for a price."

"How much do you have in mind?"

"I was thinking more along the lines of you two paying me a visit from time to time. I think we need to work on your technique if you are ever going to become accomplished robbers."

"Deal! I don't want to rush but Roderick and I have to be getting home. How about us coming by again in another couple of days to visit?"

"Deal! I'll look forward to it."

Marianne nudged him.

"Lee, are you okay? You were mumbling, ah, something about a witch."

"Whew! I just had the strangest dream. In my dream, I was robbing this old lady who lived behind my dad's store when I lived in Bend." He smiled sheepishly, continuing, "Actually, I didn't rob her. She ended up feeding me and my friend milk and cookies."

"Hmm, was your friend's name Roderick?"

"Yeah, he lived next door. How'd you know?"

"You kept mumbling his name."

"Hmm, I wonder…."

"Wonder what?"

"I was just wondering why after all of these years I'd dream about what we did to Mrs. Penrose."

"How come you set out to rob her?"

"Roderick and I thought she was a witch. Looking back, we were probably just curious."

He shook his head, staring at the wall.

"She was really nice, real lonely, though."

"What happened to her?"

"A few days after we tried to rob her, she died."

"Strange, your dream was almost like a premonition."

"A premonition, of what?"

"I doubt anyone would think of me as a witch, and hopefully I'm gonna last a little longer than a few more days, but maybe your dream was symbolic of something else. Yeah, some people believe dreams actually mean something. Course, ah, you have to figure out what it is."

"So, you think my dream had might have a special meaning?"

"Could be."
Her eyes glazed over, dulled from the distance suddenly separating them. "I guess time will only tell."

BELL-BOTTOMED TROUSERS

Once there was a sailor boy
Sailed the ocean blue
He loves his sweetheart and she loves him too
Soon they will be married and raise a family
And dress up all their kiddies in sailor dungarees
Bell bottom trousers coat of navy blue
She loves her sailor and he loves he too
–unknown

Sitting on the airplane bound for Providence, Lee's quest for a naval commission was finally about to begin. Although better than ending up in a foxhole, mustering enthusiasm was difficult.

The stewardess stopped her cart in the aisle beside his seat.

"Would you like something to drink?"

"May I have some orange juice and a cup coffee?"

"Certainly. Are you having breakfast?"

"Sure. When you're done serving, could I possibly get a pen and some paper?"

The stewardess smiled, knowingly.

"Miss her already?"

"Something like that."

I'm not sure what to make of the weekend I spent with Marianne. For someone who has supposedly met that important someone, she sure didn't act like it. No matter, seeing Sandy revived my hopes. If I play my cards right, something might just work out between us. Otherwise, why would she have the cocktail waitress tell me not to start without her? Yeah, I think I'll drop her a line. I've loved her since the first time we met.

He took a sip of coffee.

I doubt Marianne will write. Even if she did, I'm not into musical beds. Guess I'll catch a little shuteye as soon as I finish writing this letter.

After the plane landed, Lee quickly set about claiming his baggage and checking to see where the bus to Newport was parked.

Providence must not be very big. Shoot, if the airport's any indication, it can't be any bigger than Portland. Compared to O'Hare, finding my way around this airport will be a piece of cake. Hmm, there's where I claim my bags. Good, the bus is parked out in front by the curb. Only the Navy would paint a bus blue and yellow with U.S. NAVY painted on it. Suppose they're advertising?

Waiting his turn to check in, Lee observed the bus driver's deliberate manner.

He sure doesn't seem to be in any hurry. At this rate, I could be here for awhile. No matter, it'll give me time to enjoy my last moments of freedom. Before too long I'll be doing everything the Navy way. I wonder what it's going to be like? Ah, I guess it couldn't be any tougher than ROTC summer camp or going through fraternity initiation.

The driver finally reached for his bags. He smiled warmly.

"Long ways from home?"

"Yeah, I'm from Salem, Oregon."

"You don't say. Is Oregon anywhere near California?"

"Uh-huh, North. Ah, out West where all the wild Indians live."

The driver chuckled, continuing the task of loading the bus.

"Sure, now I remember. Is it anything like what you see in the movies?"

"Nah, we've been at peace with those savages for quite awhile, now."

Moments later, all loaded and ready to roll, the driver checked in his rear view mirror to insure everyone was seated.

"Next stop, Newport. Welcome aboard."

Groaning and vibrating as it pulled away from the airport, it wasn't long before the transit vehicle entered a busy corridor, settling into the outside lane.

A cadet seated beside Lee elbowed him gently, probing, "Nervous?"

"A little. How about you?"

"Not really."

"Where are you from?"

"Oregon. My name is Bob Cardon."

"Really? Me, too. I live in Salem."

"I know. I overheard your conversation with the bus driver."

"My name's Lee Grady. Ah, where did you go to school?"

"Lewis and Clark, and you?"

Lee wrinkled up his nose.

"Willamette. Can you believe it?"

"Small world."

"Yeah, I guess we're neighbors."

"For sure. Three thousand miles from Portland and this Pioneer runs into a Bearcat."

"I doubt you'll have to worry. We're teammates now. You want to play quarterback, or are you good at catching passes?"

"I'm a golfer. Well, Willamette, what are your thoughts about hooking up to do a little sightseeing when we get some free time?"

"Sounds good to me, but I've got a long list."

"Me, too. Ah, I don't know about you, but I probably won't make much of sailor. I only joined to avoid the draft."

"When I was at Willamette and involved with Air Force ROTC, I had intentions of making a career out of the military. Now I'm not so sure. Four years sounds like a long time let alone twenty years."

"Four years? Humph, if you don't complete OCS training, you're obligation is reduced to only two years."

"Really, how do you know?"

"I've got my sources."

Bob glanced out the window, announcing, "Hmm, we've arrived. So this is Newport. Jesus, will you look at those mansions? Sometime I'd like to take a tour of this place. Course, there's a few other places with a little higher priority."

"They're sure big!"

"And flat out gorgeous, just like you see in the movies."

"Um, about not completing the training, Bob, are you saying you don't intend to get your commission?"

"I need to be convinced. What about you?"

"I'm no quitter, but you've definitely got me thinking."

When the bus arrived at the entrance to the base, the guard checked the driver's credentials. Satisfied, he waved the bus through the gate.

"Take a left at the intersection. An officer will meet you in front of the Administration Building."

A moment later, the bus rolled to a stop. A Lieutenant dressed in his dress khakis awaited them. Boarding the bus, he motioned to a striped enclosure outside the bus.

"Fall into formation and put your gear over there. Make sure you have your orders out."

Lee frowned as he moved towards the exit of the bus.

"Didn't we already have them checked?"

"Get used to it. There's the right way, and there's the Navy way."

A few minutes later, after checking orders, the Lieutenant called them to attention and marched them to the barracks where they were to be quartered.

"You've got five minutes to find yourself a bunk and stow your gear on top of it. When you're done, get your butts out here and fall back into formation. Well, get a move on, girls, we don't have all day."

Inside, the gear flew, as cadet baggage found its way to the neatly aligned racks. Moments later, after they had again assembled in front of the barracks, the company officer surveyed the troops.

"Gentlemen, we're going to march over to supply and get you fitted for your uniforms."

Lee chuckled as he watched some of the cadets stumbled over themselves trying to keep in step.

"I wonder if they'll ever figure out which foot should hit the ground first. Think they'd be any good at the Houston two-step?"

"Maybe, but from what I can see, they're be a lot better at the Dallas shuffle. Have you ever seen anything so fucking pathetic?"

"Only in the movies."

After arriving at the supply depot, they were issued the clothing they would wear while at OCS. Bob frowned as he held up his shorts.

"God, I hate skivvies. The long fly may be good for some things, but...."

"Everything hangs out. Maybe it's how hanging five came about."

"Maybe for you, but I'm a six. Pioneers are well endowed. Hey, the dungarees aren't so bad. They're almost like civilian jeans."

"Yeah, but the shirts remind me of what the convicts at the Oregon State Penitentiary wear. Hey, would you look at this dress uniform. It's just like the officer's wear."

"It is. The only difference is the insignia."

"I suppose we still have to spit shine our shoes. Fuck, the armed forces must think spit and polish is the solution to everything."

Moments later, loaded with their gear, the recruits were marched back to the barracks.

"You have thirty minutes to pack your civilian clothing, stow your Navy issue and get dressed. You roll your sock and stowed 'em with the smiles pointed up. I don't want to see any frowns. Oh, from now on, you will be taking your orders in the barracks from your cadet squad leader. I want to introduce you to Cadet Charles Rankin. Cadet Rankin comes to us from the regular Navy with six years of experience as a Radioman, First Class. He'll square you away on the routine around here after you visit the chow hall. Ah, there's a lockup room in your barracks to stow your civilian gear. You can check it out when you leave base on liberty if you want to travel incognito."

After dressing and stowing their gear, they again formed up outside their barracks.

"I'll bet the next stop is the barber shop."

"Doesn't concern me."

"How so?"

"I had my hair cut by a retired Navy Chief before I left Portland."

"I doubt it'll matter, Bob."

A few minutes later, en route to the mess hall, Lee grinned as he surveyed Bob's haircut.

"Don't suppose the chief missed a few spots, do ya?"

"Humph, it'll just make it a lot easier to wash. Ah, you going with anyone back home?"

"Just corresponding."

"Good luck! This life isn't too good for relationships."

Back in the barracks, after discovering Navy food was more than adequate, Lee listened as the cadet squad leader outlined what they could expect.

"The first week we're gonna get indoctrinated. It's a fancy way of saying you're gonna get shots, visit the dentist, take a few tests, and get a little exposure to leadership training. For all of you with two left feet, we're gonna teach you how to march. We'll get down to the classroom business the second week when we're introduced to classes in seamanship, navigation, naval operations, and engineering. Any questions?"

One of the cadets asked, "When do we get to leave base? There's a few places I want to see."

"We'll get our first pass off the base in three to four weeks."

"Until then?"

"We're on the base."

"Any restrictions where we can go once we're allowed off the base?"

"Usually it has to do with distance you can easily travel in a day."

Bob turned towards Lee and mumbled, "Three weeks is a fucking eternity."

"What does it matter? We don't have transportation, anyway."

"Have some faith. I met a guy who has a car. I'll introduce you to Jim Leffords when this meeting's over so we can start getting organized."

The prison sentence passed slowly. Marching on the quad, going to chow, attending classes, nightly study sessions, and spit shining shoes soon as routine as breathing. Three letters passed through the post office, two written by him. Marianne's pen probably ran dry. Although not an immediate response, Sandy's letter pleasantly surprised him.

Hmm, she's glad to hear from me and doesn't seem to be bothered about having to wait for me to get out. Time will tell, but four years seems like an eternity. Maybe she meant it after all when she asked me not to start without her. No promises and none expected from me.

Smart! Hmm, looks like her career is moving right along. She's gonna prosecute her first case. No wonder I haven't received a letter from her before this. Sounds like she's busy as hell.

Parole! On Friday, after their last class, Lee and Bob joined Jim in the line to get their liberty passes. Excitement flowed like current from a generator as they inspected each other's uniforms.

"You look pretty sharp, Bob, ah, you too, Jim. Except for the insignia, we all look like the real deal."

"No loose threads?"

"Nope, none I can see. How does my tie look?"

Jim snorted, contemptuously mocking the activity that was taking place, "We're acting like a bunch of high school girls getting ready to go to the prom. Shit, we look fine." Smiling slyly, he quickly added, "Besides, we have more important things to worry about than our uniforms."

"Name one."

"Well, when we get to the Big Apple, we need to find a place to stay. Then there's a small matter of finding some babes."

A cadet standing in line next to them interrupted, "I don't mean to eavesdrop, but I might be able to offer a suggestion."

"Are you from New York City?"

"No, but my girlfriend is, and I go up there all the time to see her. Um, why don't you try the Trask Hotel at Broadway and 48th Street? When I stayed there, I found the price to be very reasonable, and it's right in the heart of the Times Square area. You shouldn't have any trouble connecting with the ladies, using the Trask as a home base."

"Thanks for the information. Ah, my name's Lee Grady."

"Brent Jawings. I'm guess I'm headed your way." Shrugging, he casually added, "Going up to the big city to spend the weekend with my girlfriend."

The duty officer at the gate interrupted, "If you want to stand there all day speculating what you're gonna find off the base, fine, but you're going nowhere without your liberty pass. Sign out right here on this form."

Thrusting the clipboard at Lee, he inquired, "Where're ya headed?"

"New York City."

"You should be okay. This pass only allows you to travel as far as you can go and still be back within forty-eight hours." He winked, explaining,

"Actually, since you don't have to be back until 1600 on Sunday, you have an extra four hours."

Moments later, they jumped into Jim's VW. Snickering like little schoolboys, they rambled on eagerly, anticipating the experiences awaiting them. Finally Jim squelched the excited exchange.

"Hey Guys! Ahem! Guys, I've been to the big city before. I'd be glad to serve as the tour director." Grinning, he boasted, "It's only appropriate someone from the largest state shows you the ropes."

"Forgot about Alaska, didn't you, Jim?"

"Um, excuse the oversight. I keep forgetting about the new kid on the block."

The anticipation of seeing New York City for the first time was hard to contain, but as Jim's Volkswagen emerged from the tunnel connecting New York and New Jersey, the wait was worth it. Lee's visceral sensory organs tingled as the panorama of skyscrapers and bright lights exploded like an electric storm in Oklahoma.

An apparent clog to the busy congested artery leading to the renowned hive suddenly caused all cars and trucks to slow. Poking their heads out the windows like any first-time out-of-towners, the slow creeping start-and-stop traffic allowed ample time to drink in the city's majesty.

"Look, there's the Empire State Building."

Jim slammed on his brakes to avoid hitting a car that had abruptly stopped without warning. With nothing to do but wait for the traffic to resume moving, he, too, started scanning the jagged skyline aglow with bright, twinkling lights.

"Whew! Impressive isn't it? Even an encore visit is breathtaking."

"Now that's what I call an understatement! Can you believe all the activity and God-awful noise? New York's a madhouse."

The whine of a departing subway train and the shrill shriek of one arriving blended in with the din created by the thousands of people scurrying about. Horns blared and several whistles tooted. Combined with the roar of thousands of automobile engines and every other imaginable sound converging upon the panorama of activity harshly abducted their attention. If Manhattan Island hadn't consisted of dense rock, the congestion might have caused the multitudes to experience a near catastrophic seismic experience not unlike the disappearance of Atlantis.

Lee nodded, still awed by the spectacle. "Noisy, too."

Humanity flowed steadily up and down both sides of Broadway intersected by scurrying pedestrians darting across the crosswalks, impatient daredevils choosing to ignore traffic signals. A passing cab driver, irritated by the irrational behavior, screamed an obscenity and waved a clenched fist out the window of his hack. Bob whistled as Jim stopped to wait for the stoplight to change.

"Those crazy bastards are crossing the street diagonally!"

At Broadway and 48th Streets, Jim turned right.

"This is where Brent said to go, isn't it?"

"I think so, yeah, there's the Trask Hotel up ahead."

A hotel doorman greeted them as Jim pulled up in front.

"You can't park here, sir. Go down to the end of the block, take a right, and then take the first right into the alley. The parking facility is well marked."

Jim nodded appreciatively for the information, lurching into the traffic.

"Damn, that was close. That damn black stretch limousine must think he's the only vehicle on this God-awful tangled mess. Broadway's a nightmare!"

A block later after a right turn and an entry to the alley, Jim maneuvered his '59 Beetle to a stop.

"Well, we've arrived. Talk about cramped parking. Grab your bags, and let's make tracks for the hotel lobby."

A moment later they entered the rear entrance of the Trask Hotel, navigating to the front desk. After registering, they turned towards the elevator. A bellhop stood nearby with their luggage in hand.

"What room number?"

"Ah, thanks, but we can manage."

The bellhop gave Lee an indifferent glance, obediently dropping the bags on the floor.

"The elevator's over there to your right."

"Where can we go to find a little action tonight?"

The young man gave Jim a knowing smile.

"First time in the city?"

"Not for me, but I was a lot younger when I was here last. I'm looking for a different kind of entertainment this time around."

"I'd go down to 43rd and Broadway to the Coral Ball Room. There's always a good band with plenty of attractive young ladies looking for someone to meet. However, if you are looking for something different, I could make the arrangements."

Jamming a dollar bill into the bellhop's palm, Bob interrupted, "Thanks for the information. I think we'll check out the Coral Ballroom. When should we plan on getting there?"

"I think I'd plan on getting there about eight, no later than nine. It's a popular hangout for people your age."

Bob jammed another dollar bill into the bellhop's outstretched hand, hurriedly leading the way to the elevator.

Their room on the 8th floor had two beds, a pullout and a view of the street below. A quick inspection of the bathroom confirmed the housekeeping service had provided them with plenty of towels.

"Not too bad, huh? Wouldn't pass for the Waldorf Astoria, but at least it's clean."

Lee nodded his head approvingly, muttering, "Like you've been to the Waldorf. Um, what time we leaving for the Coral Ball Room?"

"I wouldn't mind doing a little exploring right now!"

Bob smiled, sauntering to the window.

"Jim, this is New York City! No need to get in a rush." Studying the activity below, he continued, "The night's young, and believe me, the babes will wait. I doubt this place ever goes to bed. We just need to be cool and not get impatient."

"Okay, so what time do you suggest? My watch says it's a little past seven-thirty."

"How about giving me thirty minutes to catch a nap. If you and Lee want to stroll around Time's Square to see how many salutes you can get, be my guest. I'll should be ready to go in about forty-five minutes."

At eight forty-five, the trio hit the street. Decked out in their OCS uniforms, at a glance, they looked exactly like commissioned Navy officers. Quickly tiring of returning salutes from the many enlisted folks on the street, Bob stopped and turned towards the passing vehicles on Broadway.

"Fuck, you can't even take a step without having to salute. My arm's starting to wear out."

"I'm having a ball. This is more fun than an ant has at a picnic. Even the officers are saluting me."

Bob looked up the street. Appearing to be clear, he ducked his head and charged forward towards their destination.

"Come on, let's get a move on before another wave of salute-happy servicemen happen by. I know you're enjoying this, Jim, but I'm getting off this damned street before.... Hey, it looks like we're in luck. If the large group of people up ahead is going to the same place we're going, we've arrived."

Curious smiles greeted them, setting off a wave of salutes.

"Permission to come aboard, sir."

Lee acknowledged the greeting, electing to overlook the innocent taunting.

"Are you folks in line, waiting to get in?"

"No, sir. We're the greeting committee."

Bob chuckled as he pushed his way through the human barrier.

"You won't mind if we skip on by and buy an admission ticket, then, will ya?"

Inside, it was teeming with activity. The dance floor was packed, as were all of the tables surrounding the dance floor.

"Holy shit, will you look at all the good looking women in this place."

"Uh-huh, and the music's not bad, either. It's kinda soothing, yet real upbeat."

"Yeah, and I'm in a dancing mood. Well, Willamette, while you're standing here trying to figure out what to do, I'm gonna check things out."

"Bob doesn't waste any time. His scope is already up. He reminds me of a submariner on patrol."

"Where I'm from, it's called trolling."

"And where I'm from, you at least have to bait your hook."

"Maybe he uses lures. Whatcha say we grab those empty seats over there at the bar and order a couple of drinks."

"Good plan. Man, this place is alive, but I think the band's getting ready to take a break."

No sooner had the drinks arrived, than the leader of the band put down his guitar and approached the microphone.

"Freshen up those drinks and plug the jukebox with quarters. We'll going on break, but we'll be back in fifteen."

"I like this place, Jim. I think we're in for a lively evening."

"I like the ratio. There's good looking women everywhere."

"Yeah, like the auburn haired lady occupying the table over there."

"Where?"

"See the three gals at the table right in front of us, the table with the redhead sitting at it?"

"Wow! They sure are foxy looking. Think they're unattached?"

"I don't know, but every time I look the redhead's way, she smiles and makes eye contact. I'm gonna see if she's trying to send me a message. If everything works out, you might get an invitation to join me."

"Lead on, Captain. Remember I don't like redheads much."

Chuckling, Lee gently pushed his friend's arm away so he could pass.

"She's not giving you the eye, partner. You'll have to be satisfied with one of the others."

"They'll do. Um, how much time do you need?"

"If I haven't motioned for you to join me in ten minutes, you'd better start looking for some action on your own."

Nearing her the table, her eyes continued to emit signals. He nodded toward the dance floor as the band again began to play.

"Care to dance?"

She nodded, rising to join him.

"So, are we jus dancin to this tune, or are ye askin if me dance card is filled?"

"Is there any room left?"

"Depends. Aye can no write what aye don't know, now can aye?"

Lee nodded as the shapely redhead slipped her arm around his neck and looked upward into his eyes.

"Yur eyes are blue like thee sea."

"Wasn't that what I was supposed to say?"

She smiled, purring, "Ye can if ye like, but me eyes are hazel."

He took her in his arms and started dancing to the soothing melody. For the moment, her name was unimportant, as the melodic sounds and the pleasant aroma of her perfume momentarily hypnotized him. The trance didn't last long. No matter how hard he tried, dancing her was like leading an unruly child away from a candy store. Awkwardly tripping and stumbling at his every effort to guide her to the beat of the music, he finally stopped and leaned towards her.

"Relax. You're all tied up in a knot. If you'll let me, I'll do all the work. Just follow my lead and flow with the beat of the music."

She smiled weakly.

"Easy to say."

"You'll get the hang of it. Ah, what's your name?"

Her pearly white teeth flashed in the glow of the dimly lighted hall.

"Me name is Laura. I nay try to be difficult. Aye jus find it somewhat awkward and uncomfortable when aye dance with someone for thee first time. Ah, and who might ye be?"

"Ah, my name is Lee. Maybe I'm doing something wrong?"

"Nay, it's just Yanks dance a wee bit different from where aye come from."

"Do you come here often, Laura?"

Appearing not to hear his question, she probed, "Are ye an officer?"

"Not yet. I'm a cadet at the Officer Candidate's School in Newport."

"My, now that's exciting! Do ye plan on becoming an admiral?"

"Heavens no! Ahem, do you come here often."

"Oh sure! Me friends, and me, ah Lord, we come here all thee time. This is thee only place where a person can go to hear decent American dance music without giving an arm and a leg."

Stumbling, she smiled appreciatively as he steadied her.

"Ye have to excuse me dancin. Aye don't seem to be catchin on to how ye move on the floor, jus yet."

Lee chuckled, trying vainly to steady her.

"I've detected a slight accent."

"Ye noticed? Aim a wee bit Scottish, ye know. Aye been here for a year or a wee bit more. Aim from a wee town jus north of Edinburgh."

"Wow! I've never met anyone from Scotland before. How do you like the United States so far?"

"What's there not to like? Ye got everything here. Aye even like the sailors. It's grand ye have decided to fight for yur country."

"Thanks, but I'm not sure I would call it fighting for my country. I'm just doing my time to meet an obligation. Ah, Laura, why don't you tell me about your country. I would like to know what people your age like to do."

"Aye, nothing would give me more pleasure, but it will have to wait. Me feet are killin me. Tell me, are ye here with anyone?"

He nodded, looking in the direction of the bar.

"My friends are seated over there at the bar."

"Why don't ye ask them to join us at me table. Me friends are real corkers."

The newly formed sextet soon laid claim to some territory on the dance floor, using the table only as a place to catch their breath while enjoying another refreshment. Laura's awkwardness on the dance floor didn't improve, but her sense of humor and delightful accent more than compensated.

Towards the end of the evening, snuggling in his arms, Laura peered into Lee's eyes hopefully.

"Thee bewitching hour's for this place's a comin. How would ye like to go uptown to a friendly little place I know about?"

She smiled anxiously as he paused to consider her offer.

"They have decent music until four in thee morning, and the beer is a wee bit cheaper."

"So, why did you and friends come here instead going there?"

"We like to get away from thee ethnic flavor they offer up there to enjoy ye American music." Smiling, she confessed, "Aboot once a week or so."

"Sounds fine to me. Shouldn't we check with the rest of the group?"

"Aye say, nay. What we do is our own damned business, but aim not opposed to asking them to go along, if ye like."

The words had hardly tumbled out of her mouth before she had turned, walked back to the table, and plopped into her chair.

"Trudy, aye favor movin uptown for a spell. My partner hasn't had his fill of dancin yet. Are ye in favor or no?"

"Ah Laura, it sounds fine to me. How do ye suggest we get there? Aye don't fancy riding the subway at this hour."

Laura eyed Lee anxiously.

"We could all share a cab. How does that sound to ye, darlin?"

Overcome with excitement, Jim interrupted, "What a terrific idea. Not only do we have the good fortune to share an evening with three beautiful ladies, we now get to take a ride in a New York taxicab."

Outside, they hailed a cab, jamming into the yellow hack with black checkered trim. Taking the front seat with the driver and Laura, Lee watched intently as the driver reached over and turned on the meter.

"Where to? Just my luck, a cab full and only one fare."

Laura inquired, "Do ye know what Germantown is about?"

He nodded, quickly wheeling his cab into the flow of traffic while Laura settled back into Lee's arms. In moments they had escaped the traffic of Time's Square and were headed for upper Manhattan.

Lee slowly adjusted his position under the weight of Laura's petite soft body.

"Sorry, but you were starting to put my leg to sleep. What's this place in Germantown like?"

"Are ye no comfortable with this fat body sitting in yur lap?"

"You're not fat, Laura. You haven't got a spare pound on your beautiful frame. I was only trying to adjust my position before my leg started having a nightmare."

"That's a good one. Ah, ye will like it at the place in Germantown. We have some real high times up there." Smiling distantly, she added, "Aye suppose aim jus lonely for me homeland. Aye bet ye are no different. You probably miss yur home, too."

Without warning Laura kissed him.

When their lips parted, she revealed, "Aim very happy to have met ye, Lee. It's been grand sharing thee evening with ye. Aye hope we'll be able to meet up again, sometime."

At three in the morning with the activity still going strong at the ethnic establishment, Laura and her weary friends signaled surrender. As she watched Bob and Jim start to get into the cab, Laura explained to the driver where to take his passengers. She turned, and grabbed Lee's coat. Pulling him close, she put her arms around his neck.

When their lips had finally parted, she whispered, "Thank ye for a grand evening.

The cab driver growled, "Come on lady, I don't have all night."

Ignoring the cabby's suggestion, she kissed him again. When their lips had again parted she smiled and gave him a gentle shove towards the cab.

"Come back to see me when ye can stay longer. I have me own apartment, ye know." Her eyes twinkling, she suggested, "I'll bet we could find some trouble to get into, don't ye think?"

He winked at her, inquiring, "What's your idea of getting into trouble?"

"Aye guess ye have to come back for a visit so aye can show ye. By the way, aim in thee phone book."

During the trip back to the hotel Jim started pumping the driver for information.

"Where would you suggest someone new to New York might go to find some real action?"

The driver, his accent distinctly a Brooklyn flavor, inquired, "Whatcha have in mind? Action around here comes in all sizes and shapes."

"I was wondering where I could find a woman who doesn't expect a person to spend the entire evening with her before she lets them have their way with her."

The driver smiled as he looked in his rear view mirror, surveying his eager fare.

"Spend is the operative word. For a twenty, I can find you a sure thing. For a little more you can tie up the entire evening. Um, are you looking for something tonight?"

"No. I was thinking about tomorrow."

"When I getcha back to your hotel, I'll give you my phone number so you can get hold of me when the mood strikes ya. I'll need at least an hour's notice. None of this bang, bang shit."

Jim turned towards his friends, sputtering excitedly, "How do you guys feel? Do you want our friend here set you up, too?"

Bob shook his head, feigning boredom.

"Lee mentioned he wants to walk around and see the sights. While he's gone, I'll probably see what I can do to help you find someone. If we don't have any success by five o'clock, we'll give our driver a call and take him up on his offer."

Bob looked into the driver's rear view mirror to meet the gaze of the driver.

"Is five o'clock too late for you to find someone?"

"In New York it's never too late, but like I said, I'll need an hour."

When they pulled up in front of the hotel, the cabby quickly scribbled two numbers on a piece of paper and handed it to Bob.

"The best time to call me is before noon. However, if you want to wait until five o'clock, call the second number. Ask for Pete Stark."

At half past noon the next day, Lee left to scout out some of the sights he'd heard about.

Calling back over his shoulder as he started to close the door, he confirmed, "I'll probably be back between four and five. If you guys aren't too busy with all the street walkers, maybe we could take in a movie or something, ah, after we check out what the USO has to offer."

"Take your time. We'll be out nosing around for the next couple hours or so. It would be nice if to find someone for Jim so he can dip his wick. He has a real bad case of the horns."

After leaving the hotel, Lee walked briskly towards the East River until he reached Fifth Avenue.

Guess I should turn left. Ah, I think Central Park is north of here.

He stopped to check his watch.

Hmm, twelve-fifty. I have plenty of time to wander around. Guess I can slow down a bit and take in the sights. Um, the advertising offices must be up there somewhere in those tall buildings. These damn buildings are like little self-contained towns. Hmm, Tiffany's should be across the street and up a block. Ever since I saw "Breakfast at Tiffany's," I've wanted to go there.

He noticed that the people surrounding him seemed to move like the ocean current, fast and powerful. Waves of people seemed to part at the intersections, a simple mechanical device, a streetlight, causing the next wave of humanity to spill into the intersection. Change after change, the light freed another clog of people to scurry on to their next destination.

Watching all these people moving down the street reminds me of a tsunami.

At the light he dreamily observed an older gentleman accidentally bump into a very attractive lady. The lady dressed in a finely tailored suit grunted and then delivered a tirade as the man bent over to pick up the purse he had caused her to drop.

After I browse around Tiffany's, I think I'll head up to Central Park.

Three hours later, Bob and Jim greeted him as he opened the door to the hotel room.

"How was your walk? Hmm, four o'clock. I honestly thought you wouldn't be back until five or later."

"Good. It was really good."

A grin creased his face.

"Hey, Cardon, didn't you help Stranton get set up? I thought I would have the room all to myself when I got back."

"No, we struck out. I guess Jim will have to call the cabby."

"Haven't you already made the call?"

"No, we were waiting for you. Well, what all did you see?"

"Mostly, I just moseyed around, but, ah, I did see Tiffany's, Central Park, and the sights between. I even saw the Central Park zoo. I hope you weren't planning on getting the cabby to set me up. I'd like to do something, but I don't want to do any whoring around. Ah, you can chase the strays for me, huh Jim?"

"What would you like to do, then?"

"How about getting something to eat and then going to some quiet bar. We can toss down a couple and then hit the sack. What'd say, Jim?"

"Ah, I guess it's okay. We'll hit pay dirt next week. Besides, I have to get some sleep so I'll have the energy to hit the books when we get back to the base. I'm not doing too well right now, so I think it'd be a good idea to do a little cramming."

The next morning, Lee gratefully accepted a rear seat assignment for the trip back to the base. "Don't mind me, but I think I'll take a little nap."

"Two of my classes are crap. Sixty-seven in seamanship and sixty-eight in operations aren't gonna cut it. If I don't get those marks up to seventy I'm gonna wash out. Fuck, I hate seamanship."

"What'd you say? You hate seamanship?"

"Yeah, don't you? I'm a wall and floor kind of guy. All the bulkhead and deck bullshit really irritates me. It's the kind of lingo only a Pioneer can understand."

"Yeah, right. How are you doing in operations?"

"No better."

"Navigation and engineering?"

"Nothing a little study won't cure."

"Hmm, it sounds about like how Jim and I are doing. Um, I think we're all in deep shit."

"Yeah! But, ah, I'm real partial about going up to Boston next week. I've got a fraternity on the Boston University campus, so finding a place to stay won't be too tough. What'd ya think?"

"I've got a fraternity there, too. Yeah, Boston could be a kick."

Jim waved his free arm in the air, excitedly.

"Okay by me. I hear Boston's hot. It might be a good place to get laid. Besides, we'll be back in our element. Chasing women on a college campus is what I do best."

"Okay, next weekend we hit Boston. The week after we can slip up to the Maine coast. I hear it's beautiful. Sound okay?"

"Lee, I hope you and Bob want to see a little more than the Maine coast. Personally, I'd to catch a little of New Hampshire and Connecticut. We can even throw in a tour of Boston's freedom trail."

Bob grinned, shaking his head in confusion.

"Somehow that doesn't seem to fit, Jim. I thought you wanted to devote all of your time to getting laid?"

"I've been know to come up for air once in a while. By the way, we're going to have to get our civilian duds out of storage. I don't know how you feel but I seriously doubt the campus chicks at Boston U. will be excited about going out with anyone dressed in a uniform. Hey, guys, try this on for size. When we're in Boston, we'll tell everyone who asks we're taking a final tour of the country before we enter graduate school winter term."

"I don't believe you, Jim. You come up with some of the damnedest ideas. Hmm, but this time I think you've struck gold. What do ya think, Bob?"

"Not so fast, guys. How do we explain the three of us traveling together when we all graduated from different universities? Maybe they'll buy two of us traveling together, but three?"

Jim clarified, "That's easy. We'll just tell them we met last year in San Diego during spring break. We had such a good time together we vowed to tour the country together before going back to graduate school. Besides, I doubt anyone is going to dig into our past."

"Why is it you didn't say we met in Dallas, Jim?"

"First I overlook the Alaskan's and now I'm starting to forget my own state. You guys are starting to have a bad influence on me."

BOSTON BAKED BEANS

A single discernment can influence or shape the direction of a person's life. Often, in discussions about an accomplishment, people confess, "If I had the opportunity to change one thing I have done in my life, I would...."

The training classes did little to inspire Lee, Navy jargon holding the same interest as the foreign language courses he had taken. Resisting the new terminology contributed failing marks. Unlike the game of horseshoes, a leaner didn't count. Bob and Jim experienced the same fate. A percentage of common sense should have convinced them to stay on base to hit the books, but when Friday rolled around, they again hit the road.

Unlike Jim, whose erotic impulses ruled his powers of reason, Bob and Lee weren't going to Boston for a roll in the hay. Boston's historical sights, revisiting campus life, Boston's famous baked beans and a hero sandwich from the pier were far more compelling.

Located in his station of Jim's VW, Lee turned up missing in action as soon as the base disappeared from view.

I wonder why Sandy didn't write? Probably busy with some big case. Damn, I'd like to see her, be with her, but ah, damn, I can't wait until I can take some leave and spend some time with her. We've got a lot of issues to discuss. Right now, I'm kinda in limbo. There's no reason for me not to go out with other people, but.... Oh what the hell, she's probably not sitting at home, either.

"You asleep, Lee?"

"No, just daydreaming."

He stared out the window as the VW sped towards Boston. The green landscape reminded him of home. Only the long parade of mansions lining the expressway corridor was different.

"Welcome to Silicon Valley. This is where all the electronic millionaires live. We're about to pass through a stretch of wealth unsurpassed by just about any place in the country, Texas excepted."

"You Texans have a line as long as the Rio Grande, don't you?"

"Wait until I start talking about the cattle ranches back home."

"Too bad I'm not from Alaska."

"Why?"

"I'd put all the bullshit in cold storage where it belongs."

"Now, now, don't get all heated up. We need to focus on our mission."

"Spare us from the details. Um, what do you figure the homes along here cost?"

"Oh, I don't know, probably a million or more. I'd suspect they're all mortgaged to the hilt, though."

"On a larger scale, this reminds me of an area where I live. People in Salem refer to the Candalaria neighborhood as mortgage hill. By the way, unless you plan to show up at my fraternity looking like this, when do you plan to stop and change into our civilian clothing?"

"We can wait until we get to Boston."

"Good idea, Bob. I've heard of a hotel with a great restaurant. Supposedly, the prices are good, too. Surely they'll have a place for us to change."

The waiter approached their table.

"I see you've changed. On a weekend pass, huh?"

Bob nodded, looking over the menu.

"Yeah, we thought we'd enjoy the hospitality of Boston more if we were a little less conspicuous."

"I know what you mean. Ah, nothing against you guys, but the uniform kinda turns some civilians off. Have you decided what you want to order?"

"What would you recommend? I've leaning towards lobster."

Frowning, the waiter adjusted a napkin and silver setting.

"Boston's known for its baked beans. Our fried chicken and a steak is also good." Leaning towards his guests, he whispered, "Between you and me, I'd wait to try the lobster until you venture further up the coast."

"We're going up to Maine next weekend."

"I'd definitely wait, then. Any place near the ocean in New Hampshire would be my choice. Some of the best places are right on the docks. It's a lot easier on the pocketbook, as well."

After dinner, they went to the SAE house where a member greeted them in the foyer.

"Can I help you?"

"My name is Lee Grady. I'm a member of the Oregon Gamma chapter at Willamette University. My friends and I are in Boston sightseeing for the weekend. This is Bob Cardon and Jim Leffords."

"Pleased to meet you. My name is Josh Smith. I'm the house president. Just dropping in to visit the house, or are you looking for a place to stay?"

Lee grinned sheepishly, suddenly embarrassed to think that Josh might view him as a freeloader.

"Maybe a little of both. Do you have a spare room?"

"Sure. You're welcome to stay the entire weekend if you'd like."

Jim interrupted, "Perfect. We were hoping to continue our tour of the East Coast on Sunday. We'd like to do a little campus cruising until then. Um, is anything going on around here tonight?"

"Not much other than the Beta function. They're hosting a function with an independent dorm. Let's see, oh yeah, there's a football game tomorrow night, if football's your thing."

Bob grinned, offering, "I belong to Beta Theta Pi."

"You've got an in, then. The name of the president is Eric Shown. He's a good friend of mine. I could give him a call if you'd like."

Lee nodded, shedding the discomfort of appearing to be scrounging for a freebie.

"Good idea. If you don't mind showing us our room, we'll freshen up and then head on over to check it out."

"You guys are just traveling around the country?"

"We're trying to see as much of the East Coast as we can before enrolling winter term for graduate school."

"Wow! Where did you begin you tour?"

"Newport, Rhode Island. We're kinda using it as our home base so we can see the America's Cup races."

"So, where are you headed after the cup races are over?"

"Who knows? We're real flexible. Ah, Josh, we really appreciate you putting us up for the weekend."

After unpacking their gear and showering, they headed for Beta house. At the Beta house Eric greeted them.

"Josh told me about your plan to see the country before going back to school. I'm envious."

Bob explained, "Yeah, we plan to finish our tour sometime in December, Eric. We'll need a little break before we begin winter term."

"We're having a mixer with an independent dorm tonight. You guys interested?"

"Will we be intruding?"

"Not a chance. You can help us keep the unattached women occupied. Most of 'em are damned good-looking, too."

"What's the dress?"

"Come as you are, you'll fit right in. Everything starts about nine."

Bob chuckled as he playfully nudged Jim.

"I'm glad there'll be some unattached women here tonight. My friend is looking, if you know what I mean."

"He's in luck. Trust me, they'll be plenty."

"Any chance of getting fixed up?"

Bob grinned, again playfully giving Jim a little shove.

"Josh, you'll have to excuse Jim. He's on the make."

"Well then, don't forget to bring a stick to fight 'em off."

Shortly before nine, Eric took them on the grand tour. In the basement, after getting a beer at the bar, they moved into a large, rustically decorated adjacent room teeming with good-looking women.

Who in the world is the beautiful blonde over there? Holy cow, she looks exactly like the actress, ah, what's her name? Oh, yeah, Grace Kelley. She reminds me so much of her, absolutely gorgeous, a sculptured piece of art. No wonder all the guys standing around her are going ga ga. I wonder what it's gonna take to meet her? Oh, oh! Looks like someone already beat me to the punch. Huge surprise! Someone as good-looking as she is doesn't have to wait long before someone grabs her up.

Eric grinned, noticing Lee's fixation on the attractive coed across the room.

"Come on, Lee. I need a beer, and it looks like you're a little dry yourself."

At the bar, he motioned in the direction of the beauty Lee was focusing his attention on.

"Um, the blonde's name is Sara. I wouldn't be misled by what you see. The guy she's with is just a friend. He's showing her off so she can hook up with someone. Wanta meet her?"

"Damned straight. Do you know her very well?"

"Yeah, she and I went to the same high school."

"Um, so she must be a senior?"

"No, she's a year behind me."

Jim interrupted. "Bob and I are thinking about getting out of here and heading downtown to check out the bars."

"You won't want to miss the Round-up Room, then."

Eager as ever, Jim probed, "Why, is it supposed to be hot?"

"Understated! Too bad you didn't rent a hotel room."

"You're kidding."

"Shit guys, Eric was about to introduce me to the cute little blonde number standing over there by the door."

"Bad news, Lee! She's taken." Bob paused, then impatiently inquired, "You coming? We're leaving."

"If I were in your shoes, Lee, I'd go too. If you don't find anything, come on back and I'll introduce you to Sara."

"You mean if she's still here."

"It's a risk you have to take, but I doubt she's going anywhere." He winked, slyly offering, "I'll put in a good word for you, and ah, while you're at it, get some for me."

"Well, okay then. I guess I'll tag along. Um, when I get back from showing these guys the Round-up Room isn't paradise, remember I want to meet her."

"I'll see what I can do, but I wouldn't be back any later than eleven."

"It shouldn't take more than a couple of hours to scope out the filly palace."

As he started to head out the door, he paused and looked back towards their host.

"Eric, thanks for inviting us to the function tonight. Ah, anything you can do to fix me up with Sara will be appreciated."

He glanced across the room, making eye contact with the blonde beauty.

"She sure is pretty. I think I could fall in love."

The Round-up Room lived up to its title with alluring game roaming the premises in abundance. Unlike wild prey, they weren't wary, content to prowl about boldly. They appeared all too eager to form a coalition. A shapely brunette tapped Lee on the shoulder.

"Want to dance? Shame to let good music go to waste."

"Sure."

"My name's Julie. I'm from Worcester. I come here every Friday night to party. Where you from? I haven't seen you in here before."

"Just passing through. So, Julie, where's Worcester?"

"Not far from here. Have plans for the evening?"

"Depends on what you have in mind."

Smiling coyly, she motioned towards a nearby table.

"Pam, the one sitting on the left, is our driver and Jenny, the one on the right, well, Jenny's just plain Jenny."

"Are you looking to hook up?"

"Not so fast, there's another friend. Rachael's around here somewhere. How many in your group?"

"My friends Bob and Jim make it a total three."

"Hmm, four to three. Any idea how we might go about evening up the odds other than shipping one of my friends home on the bus?"

"I suppose you come as a package deal, huh?"

"Exactly! Course I would listen if you have another idea."

"I don't know anyone in Boston, and I have serious doubts about coming up with a fourth."

"I'm sure if you put your mind to it, you could come up with something."

She kissed him on the ear lobe and slowly moved her pelvic region forward, making a short, inviting thrust. Withdrawing the offering, she smiled coyly.

"Enough incentive or do you need more?"

"You've got my attention. Ah, ever consider flying solo?"

"Now you have my attention. Do you have a car?"

"If Jim lets me borrow his, I do."

"Well, why don't you ask him. I don't plan on walking."

"On second thought, I don't think so."

She frowned, a dumfounded expression enveloping her expression.

"Talk about being erratic. So, now you're all of a sudden not interested?"

"It would make things far too complicated, besides, I have other plans."

When Lee returned to the table, Jim could hardly contain his enthusiasm.

"Is she hot, or what?"

"She's looking to hook up for the evening. Interested?"

"Are you shitting me?"

"No, but she lives in Worcester. Ah, you'll have to take her home. You'll also have to rent a room if you want to take her to bed."

"Maybe we should go back to the Beta house and see what we can rustle up."

Bob chuckled, jabbing Jim playfully in the ribs.

"I thought you wanted to get laid?"

"Let's head back to campus. I'm not gonna be some horny heifer's chauffeur just so I can check out her brand. Besides, Lee wants to get back to Sara."

"You sure?"

"I can wait. Maybe Eric will have us all set up."

At two minutes before eleven, they made a beeline for the dancing area. Sara was still there. Better yet, she appeared to be alone.

"Hi."

"Well, hello, traveler. I see you didn't get branded. Didn't the Roundup room live up to its reputation?"

"I wasn't interested in going on a cattle drive."

"Eric had to leave." Blushing, she continued, "I couldn't help noticing you earlier in the evening."

"You captured my attention, too. Your date, where did he go?"

"Carl? Oh, we're just friends."

"Good."

She smiled, unsuccessfully concealing contentment with Lee's disclosure.

"Oh, why is that, pray tell?"

"I was afraid you might have something going."

"Nope, still unattached. What about you?"

"Um, would you like to dance?"

On the dance floor he bent forward, brushing her ear with his lips.

"Well, are you going to tell me?"

"Ooh, you're tickling me. Ah, tell you what?"

"I'm dancing with a complete stranger. We could start with your name."

"I'm Sara Morgan, and you?"

"My name's Lee Grady. Confession?"

"Sure."

"Eric told me your name."

"Did you ask for it?" Blushing, she continued, "You don't need to answer. I asked him your name, too. So, where are you from?"

"Oregon. It's the State everyone thinks is still inhabited by savages."

"What do you plan to study? Ah, you know, when do you plan to start graduate school." She blushed, confessing, "Oops. I wasn't supposed to know, was I?"

"No harm."

He gently brushed her cheek with his lips.

"You, I plan to study you."

"What? Oh! Um, nice line! You could be a little more original, though."

Her cutting remark abruptly derailed his locomotive.

"I meant you'd be a perfect place to start."

"Rebound quickly, don't you?"

He grinned, sheepishly shrugging his shoulders.

"Sometimes."

"I don't like to spar over half truths. I prefer to be with people who are up front with me from the very beginning."

"Sorry. I just meant you've captured my attention."

"You as well, but um, I'm a little concerned about your problems with communication. Is it difficult for you to say exactly what you mean?"

"When I'm afraid I'll mess things up."

"You're doing fine, but I'd appreciate you dropping the line."

He pulled her closer, brushing her cheek again with his lips.

"Deal."

"How long will you be staying in Boston?"

"Until Sunday. Any chance I can see you again?"

She snuggled closer, responding to his nuzzling.

"Tomorrow night?"

"I was afraid you would never ask."

"You don't lack for confidence, do you?"

"Truth?"

"How refreshing."

"Actually, when I first saw you, I knew I wanted to see you again. I was scared to death you wouldn't feel the same way."

"All you have to do is ask."

"Will you see me again?"

"There's a football game tomorrow night."

"Should I pick you up for breakfast?"

"If you tell me where and what you plan to study."

"You drive a hard bargain. Um, I'll either study psychology or go to medical school to become a psychiatrist."

"Hmm, impressive! If you'd like, you can walk me home. As for breakfast, I'll be ready to go at nine."

"So, we're spending the day together, huh?"

"Possibly. Care to give me a reason?"

"It could be fun?"

"Could be?" Wrinkling up her nose, she prodded, "By the way, Mr. Freud, we have to go before my dorm locks me out."

"Dorm? But Eric told me you were a junior."

"I went to NYU my first two years. I'm staying in the dorm until find an apartment."

"Hmm, so if the dorm locks you out, what happens then?"

"I'd have to find some place to stay. Then I'd have to meet with the disciplinary committee. A indiscretion like being late would probably be worth about two months of probation."

"Gee, I thought for something so serious, they'd lock you up and throw away the key."

"No, silly, but being on probation would seem like it. No way it's gonna happen because I have no intention of being late."

"What's the curfew tomorrow night?"

"Already thinking ahead aren't you? How does one o'clock sound?"

"Wow! On Saturday's they actually give you an extra hour?"

"Come on, wise guy. While we're standing here worrying about curfew, the minutes are ticking away."

A short walk through campus found them at her dorm.

"Man, the walk to the front door is going to be a challenge with everybody standing around making out. It reminds me of Willamette."

"Embarrassed? I didn't think anything bothered you."

He put his arms around her.

"Are you hinting?"

She blushed, suddenly finding comfort in studying the pavement.

"Could be."

"Good. I've wanted to kiss you all night."

She nodded, closing the distance between them.

"Last chance. In two more seconds I'll lose my nerve."

Their lips met. Unashamedly, she pressed her body to his, her face reddening as their lips finally parted.

"I like the way you kiss, Lee Grady. Ah, I'll see you tomorrow morning at nine."

Lee bent forward and kissed her again lightly on the lips.

"You'll be in my dreams tonight. Um, I hope you believe in like at first sight."

Saturday was like some Hollywood screenwriter's invention. After breakfast, they snooped about the Boston University campus and some of the key points of interest around Boston. Arriving at the pier in time for lunch, Sara immediately went to work on some concoction she ordered from a deli.

"This is yummy but it's really hard to eat."

Putting a sauce-covered finger in her mouth, she slowly withdrew it.

"I must look like a little pig. Good to the last lick, huh?"

"My thoughts exactly."

"Thanks. Um, the answer is yes."

"Huh? What are you talking about?"

Sara chuckled, obviously pleased that she had stumped her companion.

"How does it feel to have the shoe put on the other foot? Now you know how I felt last night."

"I'll give up if you'll tell me what you mean."

Blushing, she confessed, "Yes, I feel it is possible to fall in like at first sight."

"And are you in like?"

"Maybe."

"Do you realize how beautiful you are?"

"Sure! Why else would you be interested in me?"

"Seriously, have you ever been told you look like Grace Kelley?"

"The actress?"

"Who else?"

"Oh, sure. It happens all the time. Has anyone ever said you reminded them of Rock Hudson?"

"No."

"I didn't think so."

She giggled, poking him in the ribs.

"Gotcha!"

"Um, yes you did. Did you know you're hair shines?"

"It's the lacquer." Grinning, she revealed, "Actually, I use a conditioner."

"Whatever. I guess what I'm really trying to say is you're absolutely beautiful, and I have fallen into a total state of like with you."

"You have, have you." Blushing, she stammered, "I, I really don't know what to say. You, you've caught me off guard."

"Just say yes."

"And what might I be saying yes to, Lee?"

"Something like, you've also fallen in like with me?"

"It wasn't hard to do. You're a lot of fun to be around."

"So are you. Did you know I'm very attracted to you?"

She blushed, again finding the concrete walkway a comforting sight.

"I've always believed you start off liking someone before you fall in love with them."

"And, ah, lovers should be friends first, huh?"

"Uh-huh. Successful marriages usually start out as a friendship."

"You don't believe in love at first sight, then."

"I don't know, ah, I suppose it's possible, but don't you think falling in love usually is a result of genuinely liking someone, first?"

"I suppose."

"When I get married, I want my husband to be my best friend as well as my lover."

"Hmm. So, since we're in like, the direction is clear, right?"

"Not so fast. I'd like to be courted." Winking coyly, she added, "I won't settle for anything less than a couple of dates."

"Two dates? Wow! You believe in long engagements."

When Boston U. lined up for the kickoff, the cold autumn evening quickly gave him a taste of what New Englander's refer to as ideal football weather. Not winter cold, the brisk air never the less soon encouraged a foggy cloud to form over the twenty-five thousand fans hovering inside the stadium.

Nudging Bob, Lee stuttered, "Thank, thank God, I, I was, was able to borrow a coat from one of the guy, guys at the, the SAE house. It, it's just plain, plain cold, cold."

"Try colder than a well diggers ass, or then again, the famous key bird."

"Whatever. No, no matter how, how you describe it, the, the damned wind goes right, right through you."

Joining the fray, Jim disclosed, "In Texas it never gets this cold during football season."

"I don't why we even came. Let's head downtown, Jim, before the damn key bird shows up."

As Bob and Jim got up from their seats and started for the exit, Lee put is arm around Sara.

"Cold? Snuggle in here so I can warm you up."

"Thanks. I don't need to be asked twice. Um, what's a key bird?"

"It's a bird inhabiting North Dakota or Northern Minnesota. I think I'd better let Bob explain it to you. I wouldn't want to mess up the punch line."

"Humph, guys and your hidden meanings. Um, I could be persuaded to leave, if you want. We could always go to my favorite coffee shop. Then again, we've only endured half of this miserable game. Maybe you'd like to stay a little longer?"

In the popular but nearly vacant off-campus retreat, they selected a landing pad in a back booth.

"Hmm, you'd think this place would be more crowded."

"It usually is. The game probably have something to do with it."

"How long are we staying?"

"In a hurry to go some place? It's only ten."

"And your curfew is one o'clock?"

She smiled, nervously toying with her coffee cup.

"You're suggesting we finish our coffee and go some place a little more private, right?"

He smiled sheepishly, confessing, "The thought crossed my mind."

"My roommate gave me the keys to her car."

"Nice, but it's not exactly bluebird weather out there."

"It won't be so bad if we keep the car running and turn up the heater. We'll even have our own private orchestra. I know a place not far from the dorm where we can go. It's usually...."

"Speaking from experience?"

A few minutes later, his ego and alter ego engaged in a raging debate.

You've only known her for a few hours. Don't push things too fast.

Ah, butt out. She's not exactly pushing me away. What if she wants me to touch her?

Just don't push it. Girls don't like to be groped on the first date.

This isn't our first date.

She chuckled. "Who's winning?"

"What do you mean?"

"I wasn't born yesterday, Lee, and I'm certainly no prude. I think I can figure out what's going through your mind."

Adjusting her position in his lap, she looked into his eyes.

"How familiar does this dialogue sound?" Clearing her throat, she continued, "Should I touch her?"

"Should I let him touch me?"

"If I touch her, what will she think?"

"If I let him touch me, what will he do next?"

"How far should I try to go?"
"How far should I let him go?"
"Should I wait?"
"Do I want him to wait?"
She chuckled, smugly searching his eyes for an acknowledgment.
"Well?"
"I must really be easy to read."
"It's written all over your face. If it's any consolation, I've been thinking about the same thing."
"You have?"
She blushed, looking downward towards the floorboard.
"Maybe a little bit too much."
"Maybe we should stop. No need to do something we might regret."
She leaned forward and kissed him.
"It is tempting."
Seven minutes before curfew, she put her arm through his, snuggling close as they made their way to the entrance to the dorm.
"Thank you for the wonderful weekend. Tonight was special. No, it was so much more."
"And exactly how would you describe it."
"It was absolutely wonderful. Wanta do breakfast in the morning?"
"What time?"
"How about eight?"
"Eight sounds good. We're not leaving until ten."
She stopped at the foot of the stairs and glanced at her watch.
"We still have five minutes."
"I'm not much for making out in front of spectators."
"Me, either. I wish you didn't have to leave tomorrow."
"Me, too."
"I'm not going anywhere. I can always make room on my calendar for you."
Lee called Sara's dorm the next morning.
"I'd like to speak to Sara Morgan. Could you tell her Lee Grady is calling?"
A moment later, Sara answered, "Lee? Is something wrong?"

"Kinda. Ah, I thought I should call before I came to pick you up."

"Uh-huh! What's so important that it couldn't wait until breakfast?"

"You need to know, ah, I kinda lied about who I really am. Ah, I'm not traveling around the country, and I'm not going to graduate school."

"So, what are you doing?"

"I'm a cadet at OCS, ah, Officer Candidate's School in Newport."

"I see. Hmm, I suppose you're telling me you're in the Navy?"

"Yep. Um, I'm sorry I didn't level with you right from the start. You deserve better."

"I certainly do. Ah, remember what I said about how important it is for people to be honest and above board with me?"

"I sure do. Why do you suppose I called before I came?"

"I wish you'd told me this last night. Oh, well, it's a little late to cry over spilled milk. Ah, goodbye, Lee."

Lee put the receiver back in its cradle, staring into space.

Sara's right. It's a little late. It would have been so easy, ah, why in hell couldn't I have leveled with her right from the start? What an idiot, letting someone as special as her get away is just plain stupid. All because I was afraid to admit I'm in the Navy. Yeah, and I have to visit my instructors so I can get back on track to avoid flunking out of OCS.

STEAK AND LOBSTER

Momentum is sometime a good thing. At other times, you have to wonder if the merry-go-round will ever stop.

Lee met with his instructors the next day to determine what could be done about his failing marks in seamanship and operations.

"Chief Wengingford, the seamanship terminology is driving me nuts."

"It also seems to be scuttling your boat, sailor. At the rate you're going, you'll be out of here in another three to four weeks."

"I know. I was hoping you would offer me some suggestions."

"Seamanship isn't rocket science, sailor. If you learn the terminology, it should be a breeze."

He leaned towards Lee as though preparing to reveal a trade secret.

"Like anything new, you have to roll up your sleeves and dig in. I doubt you've over exerted yourself, huh?"

Lee met with his operations instructor next.

"Commander Manduka, I came in to find out what I can do differently. Right now, my future at OCS doesn't appear to be too bright. The operations class is killing me."

The fabled Comdr. Manduka sneered at him with visible contempt.

"You're telling me something I don't know? Doesn't look like you're doing too well in Seamanship, either."

"No, I just got through meeting with my seamanship instructor."

"What makes you think I give a shit, Grady?"

With an emphatic sweeping gesture, he dislodged a cup of coffee resting on the desk.

"Fuck, now look what you've made me do. Damn! Grady, it's bad enough I have to put up with your incompetence? Now you're starting to infect the environment. If want some advice, I'd say you'd be better off dropping out of OCS before you stain the reputation of the whole fucking fleet."

At least Chief Wengingford was nice about it, but this guy, fuck, what an asshole. No suggestions other than I should quit, huh? I know I could buckle down and do better, but this asshole takes away all my incentive. If a person had to put up with people like Manduka, why in hell would they even consider making the Navy a career?

He nodded and then rose to leave.

I know I can do it if I set my mind to it, you bastard. If I do decide to stick it out, it sure as hell won't be because of you.

After saluting, he made a hasty retreat. Snickering, he quickened his pace as he left the office.

I doubt Manduka will be bragging about going home for lunch with his wife when he sees his cronies this afternoon. They'll think he got a "noon job" all over the front of his pants.

At week's end, the traveling trio's grade status remained unchanged, but they stubbornly kept to their visitation schedule. Staring out the window as the VW wound it way up the coast of Maine, Bob shook his head in resignation. Eyes resembled glassy pools of clear water and not enough energy to even sit erect, he signaled his surrender.

"I don't know about you guys, but studying seamanship and operations wasn't gonna do me any good, anyway. Like the course we've set for God knows where, I've lost all incentive to stick it out." Nodding, he added, "Yep, this cadet is seriously considering dropping out. Ah, I wouldn't mind calling it a day, either."

Jim frowned, shaking his head in agreement.

"Yeah, what time I would've spend studying back at the base would have done me about as much good as continuing on to Portland."

The sky had started to fade, the sun hovered above the horizon. He moaned, checking his watch.

"There's no way I can drive any further. If one of you wants to drive it's okay with me, but I vote we stop and get a room for the night. I'm bushed."

"Great idea. I'm tired of riding in this damned car. Not to mention, I'm so hungry I could eat a horse."

"Careful, Texans are kinda partial to horses."

"Don't give me any more of your Texas bullshit."

Bob carefully looked at the map in his lap.

"According to the map, Casco Bay's real close. What do you think? How about stopping there?"

Lee stretched, suddenly perking up.

"I vote we stop somewhere. I don't care where, but your suggestion about stopping at Casco Bay sounds like a winner."

Ten minutes later they pulled into parking place in front of a quaint little inn located across the highway from the rocky Maine coast.

"What'd ya think, guys? Doesn't this look like something out of the movies?"

"It's rustic, all right, Bob. As long as we can get a good meal and a place to sleep without having to mortgage the farm, I'm all for it. Um, what do ya think of the view? Kinda reminds me of the Oregon coast."

"No kidding. It reminds me a little of the stretch between Lincoln City and Newport. The Maine coast is little more rugged, though."

"It's settled. Grab your bags and let's get a room. You guys can tell me all about the differences after we get a shower and something to eat."

A bulky middle aged woman greeted them at the bar.

"What can I do for you, lads?"

Bob smiled, scanning the premises.

"Your inn has a lot of personality."

"Thank you. My husband and I fell in love with it the first time we saw it. Looking for a room?"

"Something to eat, too. Do you get any action in here on Friday's?"

"Not too bad for a wide spot in the road. Ah, dinner is served in here." She motioned towards the end of the bar to an adjacent enclosure, continuing, "You can kick up your heels and listen to a little jukebox music in the adjacent room over there. This place has been known to get pretty lively from time to time."

"What's on the menu?"

"Tonight we're featuring steak and cod dinners complete with all the trimmings. Four bucks will buy all you can eat and then some."

Glancing at her watch, she mumbled, "Hmm, six-thirty. We start serving in another thirty minutes. Ah, the kitchen will be open until nine. You fellas look tired. Come on, I'll show you to your room. You'll probably want to rest a spell before dinner."

Jim grinned, obediently falling in behind her charge up the stairs.

"You read our minds."

At the top of the stairs, she stopped at the first door.

"We have two rooms. This'll be yours if it's to your liking." She motioned to her left rear, adding, "The bathroom is located behind me." Smiling, she joked, "Convenient, huh? Located right across the hall from your room. Ah, you'll need to clean up after you're done so our other guests doesn't have to wade through your mess."

The room resembled a small sleeping porch containing three standard sized beds, a four-drawer chest, and three nightstands. On each nightstand was a hand-carved lamp with nautical designs.

"Would you look at the beautiful tongue and grooved knotty pine paneling." Nodding his approval, Lee confirmed, "Paned windows, hmm, I'll bet they overlook the Ocean."

"On stormy nights, it's quite a sight."

"Not too shabby. Hmm, the door doesn't have a lock."

"Not to worry. In the five years we've owned this place, we've never lost a single guest or any of their belongings."

Bob chuckled, nervously fidgeting with something in his pocket.

"How much?"

"You lads just passing through?"

"We're on a weekend pass. We're naval cadets at OCS in Newport."

"I'm glad you told me. People serving our country always get the bargain rate, here. How does thirty dollars sound?"

"Sounds reasonable to me." Lee grinned, disclosing, "I'm reminded of the story about three traveling salesmen renting a room for thirty dollars. Course, we're not being overcharged five dollars, so you won't have to send the bellhop up to try to even up the mistake."

Nodding with a twinkle of recognition registering on her face, she responded, "Right, and you won't have to figure out what happened to the missing dollar."

"I see you've heard the story. Is it okay if I register for all of us?"

Jim interrupted, "Yeah, I'd like to take a shower and grab a quick nap."

"Not necessary, I mean you don't need to register. I know where I can find ya."

Dropping his bag on the bed, Jim inquired, "Do you want your money now?"

"Our guests usually pay up front. Saves me a lot of trouble."

"Um, are we your only guests?"

She chuckled as though unveiling a family secret.

"No, the adjacent room is rented to a couple from Portland. They come here to stay quite often to get drunk and enjoy the tranquility of our beautiful coast."

Leaning forward as she glanced over her shoulder, she whispered, "You'll have an opportunity to meet them when you come down for dinner. He's a lawyer, and she uses some very polished skills to spend his money. They're quite pair."

"How so?"

She winked, turning to leave.

"I think it's best if I let you find out for yourselves. Oh, I almost forgot. I start serving breakfast at eight in the morning. Ah, it's included in the cost of the room, of course. Ah, when you lads are rested, you come on down and enjoy the evening. The drinks don't cost much and the stupor's on the house. I'll introduce you to my husband. He should be back from picking up supplies by then."

The clock located on the wall at the back of the bar read eight-thirty when they claimed a booth for dinner.

"Wow! This place really rocks. What is there, four, maybe five couples in here?"

"Ouch! You got spoiled when we visited New York and Boston, Bob."

"Come on, Lee. This place's dead. I've seen more action on a cattle drive." Jim frowned, adding, "Getting laid is going to be a challenge."

"Ahem, what'll it be lads? We don't get a lot of unattached heifers in here. We have what you might call an atmosphere more suited to cowpokes not fresh off a cattle drive."

"I like it here. At least you can hear yourself think."

Lee yawned.

"Excuse me! The drive up here wore me out. Didn't you mention something about having Cod on the menu?"

"I sure did, but for hearty appetites, I'd recommend our twelve ounce steak. It'll melt in your mouth."

"Sounds good to me. That must be your husband behind the bar?"

She chuckled proudly, nodding in his direction

"Isn't he a handsome devil? Besides spilling drinks, he's the cook."

She turned to Bob and Jim.

"What about you two? If you're going to round up strays, you'll need something for the trail."

"We'll have the steak, too. Take away the moo, but don't cremate it. Um, could you also bring us a beer?"

Bob frowned, nodding his agreement with Jim's order.

"I'll have the same, but I need a little more than the moo removed. Texans have too much cannibal in them for my taste."

After they had finished their meals, Bob ordered another round.

"Sorry about the lack of female companionship, Jim, I guess we'll have to be satisfied listening to the music and having a few drinks."

"I'm too tired to even care. Oh, oh, get prepared, guys, the guy coming our way looks like he's had one too many."

Quickly, Lee reacted to Jim's warning, grabbing his glass of beer.

"Brace yourself, Bob, he's a big one."

A tall, burly form, unsteadily exhibiting the levy extracted by the consumption of too much alcohol, braced his bulk against the table and put his glass to his lips. After taking a short sip, he boldly plopped his heavy frame into the cushioned seat next to Bob.

"Oops, sorry 'bout that." Chuckling as he casually wiped some of his spilled drink off the table with the sleeve of his jacket, he offered, "Didn't mean to crush you. Ah, the name's John Cockran. Signal for the barkeep, will ya. I need a refill. Looks like you're a little dry, yourselves."

"Excuse me! Who invited you? If you don't mind, why don't you find somewhere else to sit your drunk ass."

Indignantly, Bob jerked away as he tended of the spillage in front of him.

Angrily, Bob scoffed, "You don't need another drink. You're way past your limit."

John took another sip from his drink and glanced at Lee, pretending to dismiss Bob's rudeness.

"I'm enjoying the hospitality of this fine inn with my beautiful wife, Judith. I was just trying to be friendly. Sorry, I didn't mean to butt in."

Bracing himself against the table, John started to rise.

"Don't rush off, John, you're welcome to join us."

Lee glanced towards the adjacent room.

"Um, your wife is welcome to join us, too, if she'd like."

John nodded, fortifying his station.

"I appreciate the offer. I'd like to buy you a drink if you don't mind. You lads staying with beer?"

"No sense in changing horses now. Ah, this is our first time in Maine. How about telling us a little about this beautiful state." Again glancing towards the adjacent room, Lee causally added, "We'd also like to hear a little about you and what you do. Um, maybe you should sit over here by Jim. I'll sit my drunk ass next to the obnoxious member of this crew."

The new seating arrangement finalized, the order from the bar arrived as John started telling them about his law practice in Portland, his wife, and some historical facts about his native state. Weaving information about geographical points of interest and special events, he performed with the adroitness of a master chef putting together a tasty souffle. While John continued to ramble on, Lee glanced towards the dimly lighted lounge, trying to identify John's wife. The survey disclosed an obvious nominee.

What a knockout! John wasn't completely honest when he mentioned she was beautiful. He forgot to mention the most important details. If I'm not looking into heaven at an angel, then I'm certainly at the gateway. Yeah, and she's also very aware of her assets.

He continued to enjoy the uncluttered view of the shapely, dark-haired enchantress, noting long, dark hair cascading down across her bared shoulders. Further inspection revealed alluring sloped extensions of her tanned shoulders supporting a dainty pale pink dress's narrow spaghetti-like straps. Her arms were folded, crisscrossed, resting on the table, cradling bountiful, alluring mounds. Obviously offering an uplifting presence to veiled eye-catching knolls, the viewer was petitioned to take more than a casual glance at her inviting cleavage.

Noting Lee's apparent interest, John suggested, "Go on and ask her to dance. She'll have to wait all evening for me to ask her."

"I don't know. What's she gonna think about a complete stranger asking her to dance?"

"What makes you think she hasn't already got her eye on you?"

John took another sip from his drink.

"For Christ's sake, man, get off your ass, and go introduce yourself. She loves to dance. Besides, you'll be able to survey the merchandise a little better over there."

"Lee, if you don't want to dance with her, I will."

Glancing menacingly at Jim, John's steely glare ripping through space until it reached its mark.

"I didn't mean you. I don't want no fuckin' fag messing with my wife."

Bob leaped to his feet, leaning towards the elder gentleman in a threatening manner.

"Knock it off, motherfucker. If you're not careful, I'll take you outside and kick your fat ass."

In a smooth, continuous motion, Lee placed his hands on the table to form a barrier between Bob and John. Glancing back and forth between them, he spoke with steel cold calmness.

"Sit down and cool off, Bob. This isn't the time or the place."

He leaned towards John, who had slumped back against the padded backrest of the booth.

"Are you okay?"

John dismissed the incident with a wave of his arm, returning to his drink.

Lee glanced at Jim, inquiring, "Everything okay?"

"Sure, go on and ask her to dance. I got the message."

"You better hurry before she gets bored and comes over here trying to convince me I've had enough."

A quick survey of Bob and Jim's demeanor convinced him the incident was over. John's infatuation with whatever was poured into his glass cinched the appraisal.

"Yeah, we wouldn't want her to cut you off, would we?"

He turned towards the table Judith occupied, and took a deep breath, trying unsuccessfully to repulse the thoughts darting through his mind.

Holy shit, she's flat out beautiful. Hmm, considering the condition John's in, if this beauty were willing, I could probably keep her busy all night. It's real tempting. Yeah, it could be flat out phenomenal getting involved with her. Too bad she's married. What the hell, I'm just gonna have a couple of dances with her and then call it a night. It's time I stop chasing every filly I run into.

"Ah, excuse me, um, your husband suggested I ask you to dance."

She smiled politely, nodding with the assurance of someone well beyond his years.

"You don't say. And what do you suggest?"

He blushed, immediately suspecting he was in over his head.

"Maybe I'd should start over."

"It might be a good idea."

"It would give me great pleasure to have the next dance."

She nodded, gratefully accepting his outstretched hand.

"Much better! I like a man who knows what he wants."

The magnificent form accompanying him to the dance floor took his breath away, more beautiful than he dared imagine. From a distance, she had radiated elegance, but up close her overpowering presence was intoxicating. Everything from her exquisite loveliness and shapely torso to the pleasantly subtle aroma of her perfume hammered at his senses, disrobing him and exposing every desire. In a dark, spacious corner of the dance floor, he slipped his arm around her slim waist. Taking her tiny right hand in his left, he deftly rotated her into position and then began moving to the jukebox's melodic strains. As the music swirled about them, their eyes met. Her thirty-something years held no importance other than assuring him the roads he yearned to travel had already entertained other visitors. To her, the eagerness of his youth promised a fiery, expedition of passionate union.

The distance between them was slight, only a momentary inconvenience.

Cautiously they eyed each other with thorough investigative tours, searching for a clue, an invitation. Finally captivated by the romantic music's suggestive mood, he drew her into close, intimate contact. Shuddering as he deliberately brushed her cheek with his lips, she offered no resistance. Again, when he pressed his cheek to hers and gently brushed her soft smooth skin with his lips, she did not protest. Their bodies united, welded at every protrusion, they moved as a single, graceful unit. Finally unable to find reason for restraint, he took her ear lobe between his lips, caressed it, finally letting it go.

"You are so beautiful. I'm having trouble breathing."

She moved her left hand from his shoulder to the back of his neck, seductively pressing her firm breasts against his chest.

"You flatter me." Slowly she whetted her lips, inquiring, "How old are you?"

"I confess to thoroughly enjoying twenty-two years. Would you mind telling me your name?"

She smiled, nodding her approval.

"Twenty-two and going on forty, I'll bet. Hmm, Judith Cockran, didn't my husband tell you?"

He grinned sheepishly, trying to conceal knowledge of what she already knew.

"He may have, but I was already too consumed with your presence to notice. Ah, my name is Lee Grady, and I'm totally captivated by your beauty."

"Trying to impress me?"

She winked, smiling seductively.

"I think we should get the age issue out of the way, first, don't you?"

"I was hoping age wouldn't be an issue."

"It doesn't have to be." Smiling as though trying to recall earlier times, she added, "Funny, it seems like only yesterday I was your age."

"Yesterday wasn't long ago, was it?"

She moved towards him until their cheeks again and took a deep, steadying breath.

"Thank you for the compliment." As his lips again press against her ear, she whispered, "Age should never come into play at a time like this.

Experience, on the other hand, is another matter. Um, I noticed you looking at me from the other room while my husband tried to ply you with liquor."

"I couldn't take my eyes off you."

"You'd like to kiss me, wouldn't you?"

An immediate warm rush of emotion enveloped him, causing a surge of sensation. Unable to conceal his physical eagerness, he pressed forward hoping their fused bodies would disguise the reaction.

"Very much."

He glanced at the top of her dress, an unfastened button inviting further investigation. As he searched for appropriate words to defend his petition, his eyes came to rest on her copious, partially exposed breasts.

Fuck me, they're huge! Would I love to get my hands on them, or not? Damn, I wonder what she just said?

She smiled innocently, trying to conceal the elation she was feeling from his flattering survey.

"Do you like what you see?"

"They're magnificent."

She nodded, pressing them firmer against his chest.

"Kiss me? We can figure out our next move, afterwards."

Breathing irregularly with anticipation he leaned forward and greeted her soft responsive lips. Whimpering as she eagerly accepted his probing tongue, her chest heaved upward. After a passionate exchange, she inhaled deeply while casually glancing around the room. She smiled, after a quick survey of the premises.

"Seems we have the dance floor all to ourselves."

He joined her survey. At his distant table in the adjacent room, everyone was too consumed with drink and a lively discussion to notice what was taking place on the dance floor. The other patrons, only a handful near where they stood, were also oblivious. Even the owners of the inn appeared to be busy with far more important duties than to review Judith and his performance.

"Yeah, and what's more, I doubt anyone gives a hot damn what we're doing, either."

She smiled. Pressing her cheek to his, she took little notice when the melody on the jukebox ended.

"You've made this evening very special."

He chuckled. Shaking his head in disbelief, he casually expanded the distance between them.

"Why would your husband allow a stranger entertain his wife?"

"It hasn't seemed to bother you up to this point. Let's go over next to the wall where it's darker and we won't be so conspicuous."

"Are you sure?"

"Unless you've got a better plan. Come on, nobody will be able to see a thing. Ah, for your information, when John gets drunk he always looks for somebody to entertain me so he can continue drinking."

A new melody began. Disinterested, she put her arms around him and pressed her pelvic region forward in a petition for closer union.

"How's that?"

"Good, damned good. Oh, fuck."

Without warning, guarding against such a reaction impossible, a sudden promotion of a noticeable urgency returned. Greeting her forward thrust, he slowly ran his hand across her soft, firm buttocks as he met her lustful gaze.

She gasped, "Yeah, it's just fine. Oh, yes, it's just fine."

Panting, she thrust her hips forward again to ensure firmer contact. Slowly withdrawing, she again thrust her hips forward. Inhaling deeply, she gasped as his veiled offering surged against her pubic region.

"Oh, yes. The big dog wants to prowl. Shall we let him out to play?"

Bracing himself against the wall, he protested, "Jesus, Judith, I didn't think you were serious. Somebody's gonna see what we're doing."

"Not here they won't. Bend your knees a little."

"I don't know. I'm not sure this is a very good idea."

"Do you know how much I want you?"

"I want you, too, but this isn't a very good place. Somebody's bound to see."

She moaned, her glassy gaze searching his longingly. Passionately she greeted his partially open mouth, moving her veiled harbor into position against his surging mass until only her panties prevented complete penetration. Again she pressed her veiled harbor against his surging mass.

"Bend your knees a little more. I left my ladder at home."

"How's that?"

"Perfect. I'll move my panties to the side so you can put it in."

"I still don't think this is a very good idea."

Gasping she whispered, "You will when you're inside me."

The disk on the jukebox stopped. He shook his head, breaking from the passionate exchange. Quickly zipping his trousers and making a few adjustments, he turned towards a table near the bar.

"I think we need to break up the party before it gets completely out of hand."

"And then what happens?"

"I need a drink."

"If you won't fuck me here, then take me upstairs to your room. My husband's about to ask you to spend the night with me, anyway. He always does. Why don't you save him the embarrassment of asking?"

"Well, he can ask all he wants, but I'm not available. Fuck, Judith, you're a married lady."

"But my husband doesn't try to satisfy me any more."

"I'm sorry, but what do you expect me to do?"

"I need to be loved. I want you."

"You say he doesn't try to satisfy you. I don't understand."

"All he does is drink. We haven't made love in, what, well, it seems like forever."

She frowned, shrugging in resignation.

"When he gets drunk, which is often, he offers me to anyone he thinks will satisfy me because it gives him more time to drink."

"I'll never understand why he doesn't make love to you. Normally, I'd kill for the opportunity, but it still doesn't change anything. I'm not gonna get involved."

"He doesn't offer me to everyone."

"Should I be flattered?"

"I really like you. We would be making love not just fucking."

A deafening silence embraced them.

After several moments of studying her, he continued, "Ah, you're the most beautiful and exciting women I've ever met. It would be so easy to satisfy my desires if I didn't think you stood for something more than a quick roll in the hay."

Frowning, she inquired, "Are you uncomfortable about how my husband will react?"

"The thought crossed my mind. Um, I'm leaving."

At the table in the other room, Bob and Jim appeared to be totally wasted, John somehow avoiding a coma. By his definition, he was still in prime condition. Through bleary eyes, a look of curiosity slowly spread across his face as Lee leaned on the table.

"So, how was she? Do you want to fuck her?"

"I'm going to call it an evening."

John took a gulp of his mixed drink.

"I'm in no condition to do justice to her. Why don't you go upstairs and do whatever it takes to make her happy? I'll keep these lads busy until I pass out."

"She's not yours to offer, but thanks for the offer. Um, whether you know it or not, she's a very special lady. If you don't mind, why don't you just tuck my friends in? They can sleep it off here. We have a lot of miles to cover tomorrow and I'd like to get some sleep."

As Lee started to ascend the stairs, the innkeeper's wife approached.

"I told you they were quite a pair. I'm glad you didn't promise anything foolish, anything you would regret in the bright light of day."

"I really feel sorry for her."

"We all have our burdens to bear."

"Keep an eye on my friends, will you?"

"Don't worry! When John passes out, I'll tuck them in for the night. By the way, where's John's wife?"

Lee looked around and shrugged.

"I don't have a clue. The last time I saw her she was in the other room. Maybe she went to bed. Um, I enjoyed the evening, but I'm going to hit the hay."

Later, responding sleepily to the tender bussing of soft lips on his cheek, Lee grunted. A faint scent of perfume invaded his nostrils and then soft, velvety tentacles began to caress his sensorial domain.

Rolling over to greet the shapely invader, he whispered, "What in the hell are you doing here? I thought you went to bed."

She kissed him again on the cheek.

"I just finished helping Edith put John and your friends to bed. Relax, there's nothing to worry about, they're all tucked in for the night."

"But, ah, I thought I told you I wasn't interested."

"I thought in a different environment you might change you mind."

She slipped her hand down the front of his shorts.

"Hmm, I was right. Why don't we let this guy out to play?"

A storm, appearing unexpectedly, howled noisily outside. Lashing out at the quaint little inn, its gale-like force caused the windows of the upstairs room to rattle as shadows danced on the ceiling above. Suddenly a loose shutter banged against the window frame as the bed whined and creaked under the strain of frantic, mutual exertion. Lost in the blackness, whimpers of contentment whispered to the walls of the chamber as the din of the storm continued to rage outside.

The next morning, two travelers sporting bad hangovers and another looking like some horse had decided to hitch a ride, waited for breakfast to be served. The innkeeper's wife chuckled as she laid the breakfast plates in front of them.

"You fellas look like the sad result of a failed stretch run at the Kentucky Derby. Then again, maybe you look like you've been ridden hard and put away wet. It's totally understandable for a guy like old John, but I had higher expectations of you. Um, I couldn't help but overhear your discussion about the next leg of your trip. Head back towards the base, huh?"

Lee nodded, feigning interest in the food before him.

"What do you think about our plan to go to Portsmouth on our way through New Hampshire?"

Jim interrupted, "We want to find the best place to stop for lunch."

"Not far from Portsmouth is a nice little spot called Seabrook. On the wharf, you'll be able to tie into some of the best lobster and steamed clams in these parts. For five dollars, you'll get all you can eat. It's not too much for your budget, is it?"

Lee smiled as he pushed his plate of food away.

"No, I think we can handle it. By the way, breakfast was great. I'm just not hungry."

Bob grinned, accepting Lee's plate gratefully.

"Don't mind if I do. I'm starved. Speaking of which, how was Judith?"

The innkeeper's wife shook her head, glaring at Lee.

"The poor soul had such a time with her husband last night. She came down shortly after you went up to bed and helped me put him to bed."

Lee yawned.

"And?"

"She said something about going back up to bed."

"Do John and Judith come this every weekend?"

"In the five years my husband and I have owned this place, I doubt they've missed more than a couple of weekends. Boys, I hope you aren't mad. When you fellas passed out I figured you might as well sleep it off down here. It allowed your friend here to have the room all to himself."

She winked knowingly at Lee.

"The storm didn't interrupt your sleep, did it?"

"Nah, outside of the shutters banging once and awhile, I slept like a baby. Man, when the wind blows here, it really howls."

"That it does."

She smiled, ever determined to make her point.

"Judith is an amazing woman. She looked as fresh as a daisy when she left this morning? She sure can hold her men."

Blushing, Lee offered, "I suppose she gets lots of practice?"

"She's never struck out since I've known her."

Lee smiled as he closed his eyes. Located in his usual station of Jim's VW, he had some things to mull over.

This weekend has been something else. Our stay at the inn was a real eye opener, but I'm really looking forward to trying some lobster. I suppose that means it's time to move onward to Lexington. When we get to Boston, I'd like to see the Freedom Trail, if it doesn't take too long. Sara told me not to miss the Old North Church, Paul Revere's house, and Faneuil Hall. I may even find time to see Old Ironside. When I was hanging out with Sara last weekend, we were too busy to see any of the historical sites. I need to stop chasing every good-looking skirt. It's time to slow down a little bit and take some time to smell the roses. If I'm not careful, I'm going to be chasing new experiences so hard I won't recognize an important one when I come face to face with it. Fuck! I don't know what's the matter with me anyway. Sandy's the one

I want, so what am I doing chasing after all these other fillies? It's time to start being honest with myself. Yeah, and it's high time I got a letter off to Sandy before she thinks I've dropped off the face of the Earth.

SCRAMBLED EGGS

When a person stops to reflect upon the image in the mirror, taking a moment to reassess the direction they're taking, the real person is often revealed.

The three travelers met Wednesday evening after dinner.

"Sure, I'd like to leave the base and go to the capital with you, Jim, but I'm gonna stay on base and do some serious cramming."

"I don't believe it. You're actually gonna apply for reinstatement, aren't you?"

"Yep, but then I'm gonna tell 'em what they can do with their damn commission. I just want it to be my decision."

"I don't know why you'd go to all that trouble, Bob. The sooner I get out of this fuckin' Donald Duck club, the better I'll like it. What about you, Lee?"

"I'm meeting with Lieutenant Crowder Thursday. I'm gonna apply to roll back, but I have to admit, two years instead of four sounds pretty damned good right now. I just need a little more time to think it over."

Lee met with the company commander the next day.

"So, tell me why you want to apply for reinstatement to the program, Grady."

"I'm no quitter. I know my grades in seamanship and operations aren't very good, but I'm sure I can make it if get a second chance."

"You know what rolling back means to your time on active duty, don't you?"

"Yes sir. I'll be obligated for almost four and a half years." He grinned slyly, continuing, "That all hinges on my application being accepted."

Lieutenant Crowder smiled.

"You have it all figured out, I see. Well, I don't see any reason not to recommend you for reinstatement. You'll meet with the Officer Review Board sometime next week."

"Thank you, sir. I really appreciate it."

"Just so you know, ah, sometimes the officers on the review board can be real bastards. Be prepared to eat a little crow when you meet with them. Usually the board accepts my recommendation, but just in case they don't, you should be prepared to go to boot camp at Great Lakes."

Later, Brent Jawings approached Lee in the chow hall.

"You've been hitting the trail pretty hard lately. Any plans for the weekend?"

"Looks like I'll be staying on the base. Since I applied to roll back, It might be best to stay aboard and do a little studying."

"Hmm, too bad. I'm going up to Cape Cod for the weekend. I was hoping you'd join me."

"Don't you usually go to New York to see your girl?"

"This weekend, my fiancée is meeting me at the Cape. She's bringing a friend. Either I find a date for Trish, or I'll have to ride herd on both her and Diedra. Ever been to the Cape?"

"No, I'd really like to go, but, ah, I just can't afford it."

"What if I told you Diedra's picking up the tab?"

"I'd say your proposal's sounding better all the time."

"Well, then, here's the clincher. Diedra's friend, ah, she's a real looker."

On Friday, after again securing a liberty pass, Lee settled into the front seat of Brent's car, folded his arms across his chest, and closed his eyes.

"Well, Brent, don't you think it's about time to fill me in on the agenda for the weekend?"

"We're gonna meet the girls at the inn where Diedra made reservations. Ah, as you'll discover, she's the coordinator. I usually just sit back and let her lead the way."

"How long have you known her?"

"Six years. Yep, we've been going together since our freshman year at Brown. Um, we're getting married right after I get my commission."

"You didn't graduate last year?"

"No, two years ago. We spent some time touring Europe."

"Hmm, how well do you know Trish?"

"Like a sister. Anything you want to know about her?"

"Well, for starters, is she going with anyone?"

Brent smiled, glancing across the seat at his companion.

"She's unattached. Ah, it might be good for you to know she's not too keen on guys coming on to her right from the get go. She likes to get to know them before diving into bed. If you just let things develop, you find she's a kick in the pants, and, ah, I meant it when I told you she's damn good looking."

"Uh-huh. Just another one of those wholesome gals all her friends like because she doesn't take the spotlight away from 'em."

"Seriously, she's a fox, an absolutely beauty. I'll deny it if you ever open your mouth about this, but Trish is much better looking than Diedra, and Diedra is flat out beautiful. Trust me, you won't be disappointed."

"My lips are sealed. I don't kiss and tell."

"Good. I'm not in the mood to pucker up, just yet. Ah, Trish only agreed to come under the condition you would be someone she wouldn't regret meeting. She hasn't had very good luck with the guy scene, lately. Um, I think she's a little gun shy."

Brent smiled sheepishly, uneasily shifting in his seat.

"By the way, I guess it's best to tell you Diedra and I are gonna room together."

"How cozy. What about Trish and me?"

"Well, since Diedra only rented two rooms, I guess you and Trish are going to be roommates, too."

"And of course Trish knows all about the arrangement, right?"

"Not yet. Um, Diedra's probably telling just about now. There shouldn't be a problem. The room has two beds."

"Good to hear. You're really putting her in an awkward position, not to mention me. Fuck! This whole idea sounds pretty screwy to me. I dread to think what's going through Trish's mind right about now."

"Trish is a good sport. She'll handle it just fine."

"Maybe so, but if you wanted to spend the weekend with your fiancée so bad, why'd you invite Trish and me along?"

"Diedra and Trish already had the weekend planned, a little get away for the girls, if you know what I mean."

He smiled sheepishly, again shifting uneasily in his seat.

"I sort of invited myself along. Ah, Diedra liked the idea, but she told me I would have to come up with someone to keep Trish busy."

"Ah ha! So, I'm nothing more than a girlfriend sitter? Is there anything else I need to know?"

"Nope, I think I've just about covered it. How about it, will you help me out?"

"Like I have a choice? If I say yes, I'll have a starring role in a tricky, deceptive plot." Snorting, he protested, "If I say no, I turn out to be an asshole. You're sure Diedra is filling Trish in on all the sorted details? I'm not sure surprising her is the best course of action."

Brent grinned, pounding his hands enthusiastically on the steering wheel.

"You're gonna help me out, aren't you."

"I'm not pleased about the arrangements, but I guess you can count on me to do my part. I sure hope Trish handles this cozy little slumber party okay. I don't relish walking on eggs all weekend."

"Thanks, Lee. I owe you."

"You'd better believe it. Well, you might as well tell me a little about your fiancée. It's always nice to know a little about all the players."

"Diedra's family is very wealthy. Her dad's a Wall Street broker."

"She's into all that society shit, then."

"Yeah. In fact, social status is very important to her."

"I'm not into the society scene, but it sounds like you've landed a pretty big fish, no disrespect intended. So, is Trish the poor sister?"

"Not at all. Let's just say her family isn't hurting. Um, not to change the subject, but you're really planning to roll back?"

"I'm considering it. I applied for reinstatement."

"Me, too. What was your stumbling block?"

"Seamanship and operations. You're definitely rolling back?"

"Yep!"

"No chance you'll walk away?"

"Not in this lifetime." Frowning, he continued, "I'm getting lots of pressure from Diedra. Her father's a retired Navy Captain, so I guess you know what I'm up against?"

"Yeah. Thankfully I don't have any pressure on me."

"Oh, there's the inn."

Brent checked his watch.

"Wow! We really made good time. I'm not so sure they're even here yet. I'll check in the office and see if they've checked in yet."

A moment later Brent exited the office, flashing a piece of paper.

"We're supposed to meet them at the restaurant across the street."

The restaurant was nearly empty except for a man and his wife occupying a booth overlooking the main street and a mother with her two children seated in a booth nearby. Two other guests occupied a corner table at the opposite side of the dining area. They were engaged in an animated, emotional debate. Appearing to be at odds, their Cokes were untouched.

Brent grinned, trying to make light of the incident.

"Want to referee?"

The blonde must be Trish. Damn, she's gorgeous! It also looks like she's a spunky little number. Yeah, kinda like a spirited filly just before it gets saddled for a high stakes race.

The brunette looked up as they reached the table, a broad smile immediately enveloping her face.

"You sure made good time."

Rising, she grabbed Brent and planted a passionate kiss on his lips.

Still smoldering, the other combatant shook her head with disgust as she toyed with a strand of her long blonde hair.

Out of the corner of her mouth, she hissed, "I hope they come up for air soon. In case you haven't figured it out, I'm Trish."

"At a time like this, maybe a song or a funny story is in order."

"Anything would be better than what I've had to endure the past few minutes. You must be Lee Grady. Um, your biography precedes you. Why don't you sit down and take a load off?"

Breaking from their embrace, Diedra made a quick adjustment to the blue sweater she was wearing as she glanced back and forth between Trish and Lee.

"Oh my God, in all my excitement to see Brent, I completely forgot my manners. Apparently no harm's been done because it looks like you've already met. Lee, just so you know, Trish is my best friend in the entire world. We've never met, but I already feel like I know you."

"Hi, Brent. How was the trip?"

"Real good, Trish, and yours?"

"Ah, revealing. Well, don't just stand there. Sit down. Diedra and I already ordered. I'll get the waitress over here so you guys can order."

On cue, the waitress approached their table.

"What will it be?"

"Ah, I'll have a Budweiser."

Trish nodded her approval.

"Um, why don't you make it two."

"Hmm, I guess Brent and I'll take on the Cokes. Ah, does anyone want anything to eat? It's on me."

Trish sneered, contempt written all over her face.

"Uh-huh. This whole weekend's on you."

"Ooh, I detect an Arctic breeze. Um, Diedra, maybe you and I should check into the room and start unpacking. I think Lee and Trish need some time to be alone."

Trish smiled coldly.

"Really! I think we'll find the time."

Smiling weakly, Diedra rose from the table, placing a dollar bill under the ashtray and thrust the keys at Trish.

"Here, you'll need these." She turned towards the exit, calling back over her shoulder, "Come on, Brent, I want to change before dinner. Um, we'll give you a call to work out the details for dinner."

"Lee, do you want me to put your bag in the lobby?"

"Sure. Tell the desk clerk I'll be along shortly."

Watching Brent and Diedra exit the restaurant, Trish methodically fingered the key to their room. Lips pursed, it was apparent a slow burn was under way.

"Fuckin' bitch! I swear Diedra makes me so angry I could spit." Sneering contemptuously, she snapped, "This must be your lucky day. You've managed to be assigned to a real bitchy roommate. Hurry up and finish your beer. I want to get out of here."

"If it's any consolation, I found out during the drive up here."

"Don't go there."

When they reached the lobby, she walked directly to the front desk clerk.

"Any messages for me?"

The desk clerk inquired, "What room?"

"Room 23."

"Sorry, no messages, Mrs. Grady." Smiling politely, he revealed, "Oh, I have Mr. Grady's bag. Mr. Jawings indicated you'd be by to pick it up."

"Well, honey, thank the man. Wasn't it nice of him to take care of your bag?"

"Ah, yeah, thank you very much."

"See, I told you we didn't have anything to worry about. The kids are fine. Guess the sitter has everything under control." She winked, continuing, "Come on sweetheart, let's go to our room. We still have an hour before we have to meet our friends for dinner."

Lee grinned, deciding to assume his role in the play.

"Sure you don't want to call to check on the kids?"

"It isn't necessary. If anything comes up, the sitter will let us know. We came up here to play, so I think it's about time we started, don't you?"

In the sanctuary of their room, Trish unloaded, her rapier-like incursions surely motivated by the boiling ambiguity she was experiencing.

"Hear me, and I strongly suggest you listen to me real good. If necessary, take notes because if you fuck up, I will make your life pure hell. Defiantly placing her hands on her hips, she continued, "I won't do anything to disrupt Diedra and Brent's weekend, but if you think you have a license to play house, think again!"

She pointed dramatically towards the beds.

"Your bed, my bed, you sleep in yours, and I'll sleep in mine. Clear?"

Nodding, his assent, he mumbled, "Perfectly."

"I'll play the role Diedra cast for me and I hope you'll do the same, but inside this room, if you so much as move a finger to touch me, I'll cut the fucking thing off."

"Hey, hey, calm down. This arrangement wasn't my idea."

"Humph. We also need to take care of the bathroom schedule. I sure as hell don't plan to put on a girlie show for you. If you're into peeping, go down to the stand and buy a magazine with a centerfold."

"Whoa, wait just a minute! I'm not any happier about this arrangement than you are, but it's not gonna do either of us any good to rant and rave about it. I got the message, so stop railing on me."

She snorted, jerking her shoulders forward as she threw her head back. Defiantly, she again placed her hands on her hips. Her exquisite face embraced an angry sneer as her lips quivered and tears began to trickle down her cheeks.

"I'm so fucking mad at Diedra I could spit. What she pulled on us is totally unforgivable."

Her insolent pose slowly disappeared, replaced by a sign of mild embarrassment.

"I know it isn't your fault. It's just I'm so fucking frustrated. Ah, in a hundred years, I would have never anticipated she would put me in this situation."

"I almost backed out."

"Why didn't you?"

"Brent didn't tell me until we were on the road."

"You poor guy. You've been thrown into a hopeless situation with an impossible bitch." Chuckling, she added, "An explosive one at that, ah, in case you haven't guessed."

"You don't make too much noise when you explode, though."

Smiling at his attempt at humor, she clarified, "I just wanted you to know the rules. I'm sure we can have fun as long as we understand how the game is to be played."

"Um, I promise to behave like a perfect gentleman. By the way, I took copious notes during the lecture."

"Funny. I wonder if the story they told the desk clerk was as good as ours. I'll bet they didn't mention having kids."

"The sitter routine was a nice touch."

"Thanks. Ah, you're much better looking than I had hoped."

His cheeks immediately reddened.

"Thanks. You stole my line. If you're interested, you did okay on Brent's report card."

"Just okay? He didn't even try to build me up, a little?"

"He told me I wouldn't be disappointed. I'd say his report was pretty accurate."

She blushed, finally rendered speechless.

"Um, the report on you was pretty good, too."

"You're even prettier when you're mad."

"Could we forget about my outburst? I don't usually use language like, ah, I don't usually swear. I especially don't use the 'F' word."

"Apology accepted. I guess I'll start putting my stuff away. Do you mind if I use the dresser?"

"No, go ahead. I'm gonna operate out of my suitcase."

Hmm, if you darken her hair and take away a couple of inches from her height, you couldn't convince me I wasn't in the presence of Diana Bodia. She even has the same kind of emotional outbursts, combustible, maybe even a little bit hazardous.

Trish began to giggle.

"What's so funny?"

Snorting, she probed, "I was wondering if you snore?"

"I don't listen to myself when I sleep. What about you?"

Snorting again, she confessed, "I don't listen to myself, either."

"You sound like a little pig when you snort." Smiling affectionately, he teased, "You're an awfully pretty little pig, though."

She blushed, busying herself with her luggage

"Yeah, yeah, we'll see what your thoughts are after you get to know me a little better."

"It could turn out to be fun. You'll have to admit, the room arrangements are sort of humorous."

"I know. Ah, some guys would die for the opportunity to shack up like this. Of course we're gonna have a different kind of fun than Diedra and Brent."

The piercing sound of the phone interrupted.

"I'll get it. It's probably for me, anyway."

She grunted as she reached for the phone.

"Hello? Okay if you say so, but we're hungry. Well, okay. Ah. I guess we'll go on ahead, then."

She stared intently, finally nodding.

"Got it! Huh? No, silly, I'll find it." Chuckling as she hung up the phone, she bluntly announced, "They'll join us after they get reacquainted. You are hungry, aren't you?"

"Starved."

"Well, what are we waiting for? Let's get ready and go."

"Where are we going?"

"Diedra says there's a really neat pub down the street."

"This pub, ah, does it serve the usual?"

Grinning, she revealed, "Yeah, lots of beer. Supposedly, they also have the best broasted chicken, homemade French fries, and coleslaw in the world."

"Sounds terrific. I'm so hungry I could eat a horse."

"You might try calming your digestive juices. I'd hate to see some poor nag bite the dust tonight."

She turned towards the bathroom.

"Just so you know, ah, I'm no prude, but when I decide to give it up, well...."

"They're engaged. What's the big deal, unless you think people should wait until they get married?"

"I'm just not into broadcasting intimate involvement. I'm also not real pleased they involved us in their little excuse to play house. If Diedra had told me she wanted to be with Brent this weekend, I would have backed away in a heartbeat."

"And miss the opportunity to meet me?"

"You have a point. Um, I'm sorry I've been so difficult to be around."

She smiled sheepishly, opening the door to the bathroom.

"I shouldn't have taken my frustrations out on you like I did."

"Just don't torture me all weekend, okay?"

"Deal. About our visit to the pub, ah, Diedra told me to put it on her tab. Did you know she plans to pay for everything this weekend?"

"Brent mentioned something about it."

"Yeah, well, the more you're around Diedra, the more you'll come to realize she's really into putting on a show."

"Well, I appreciate her gesture because I wouldn't have been able to afford this weekend, otherwise."

"Whatever. Um, I've got dibs on the bathroom."

The activity in the pub was starting to pick up when they arrived.

"Good, they have pool tables in the back room. Do you play?"

"I've taken a stab at it, but I'm not very good."

The waitress interrupted, "What'll it be?"

"Two orders of broasted chicken and fries. We'd also like a pitcher."

"We have Budweiser on tap."

"Perfect. Would you bring the beer to the pool table? I'm going to show my wife how the game is supposed to be played."

"Ouch. No problem with this guy's ego, huh? Ah, would you mind starting a tab for us? I think we'll be here for awhile. When he's through showing off, I'm going to introduce him to my game."

When they had finished their meal, Trish eyed the adjacent room.

"Okay smart guy, order us another round. I'll meet you in there at the dart board." Grinning confidently, she warned, "You'd best look out! I plan to get to get even with you for the brutal whipping you gave me at the pool table."

A pitcher later, Trish chuckled as she extracted her dart from the bull's eye.

"Hmm, twenty-one, didn't I tell you?"

"Yes, yes, you did. Want another beer? Oops, maybe I'd better order a couple of extra glasses. Looks like Brent and Diedra have arrived."

"It's about time! If they had waited any longer, we would have drained the keg."

She frowned as they approached.

"Where have you guys been? We've already eaten and nearly drank this place out of business."

"Sorry we're late. Hope you'll stay to have a couple with us."

"It's kinda late." Trish turned towards Lee petitioning, "I don't want another beer. I'm ready to call it an evening. How about you?"

"Ah, sorry guys, but Trish says she's had enough. I think we'll head back to the room."

Diedra frowned, protesting, "Ah, don't rush off. The nights young."

Lee grinned, rising to leave.

"We're working at burying the hatchet. Besides, we plan to get up early to hit the beach. What would be a good place to go for breakfast?"

"They have a Friendly's here."

Nodding her approval, Trish confessed, "My favorite. Um, would you like to join us?"

"We'll try. If not, try to save us a place on the beach, will ya?"

The Brown University graduate had warmed up considerably since their introduction. Nearing the inn, Lee casually took her hand.

"Brent told me you weren't going with anyone."

"After the start we got off to, you probably think it's because I'm so difficult."

"Explosive maybe. No, I totally understand where you were coming from."

"It could be I'm not going with anyone because I haven't found Mr. Right. Then again, maybe I want to continue traveling for a while longer."

She frowned as she released his hand to get the key to the door.

"Actually, it's because I haven't been able to find anyone Daddy thinks is good enough for me. What about you?"

"Ah, the Navy isn't exactly ideal for establishing a relationship. I should get to ride a lot of waves, though."

"The watery type I hope. Um, are you looking?"

"Isn't everyone?"

"If might help if you gave up playing the field."

"And give up my freedom? Truthfully, looking for a different girl in every port isn't too appealing."

"You could dangerous."

"Not really. I'm only twenty-two. I've still got plenty of time before I start thinking about settling down. How about you?"

"A little too old for you. Don't you know you're not supposed to ask a girl her age? To an older women, it can be quite intimidating."

"You can't be much older than I am."

"Is two years too much?"

"It all depends on how many hearts you've broken. How about it Trish, whose heart have you broken, recently?"

"None, and I have no intention of breaking yours, either. Course I may change my mind if you're set on taking the first shower."

She shoved the key at him.

"Here, I'll let you do the honors."

"Only if you tell me what you're wearing to bed so I can fantasize while you're showering."

"Now I know why I decided to wear pajamas. What about you?"

"I'm a tee shirt and shorts kind of guy. You may want to shut your eyes until I'm in bed."

The new roommates awoke early. Rolling onto her side, she chuckled as Lee rubbed the sleep from his eyes.

"Are you hungry?"

"Starved. What time is it?"

"Seven. It's about time you take me to breakfast, don't you think?"

"Sure. Ah, I didn't hear you snore."

"How could you. You talked about some gal all night."

"Sandy?"

His face paled as she nodded.

"A girl I know back home. We write to each other once in awhile, but for some reason she's stopped writing not long ago. Yeah, well, anyway, I'm springing for breakfast."

"No way, I'm treating you to a Friendly's specialty. Think we should wear our beach garb? We won't have to come back here, if we do."

"Good idea. Ah, anything out of the ordinary I need to know about Friendly's? They do serve regular food, don't they?"

"Is an omelet, hash browns, and toast with a little ice cream on the side regular enough for ya?"

"Ice cream? I've never had ice cream for breakfast, before."

"You don't know what you've missed." Grinning, she chided, "What's the matter, afraid? There's always a first time for everything."

With the opening of the curtains, Saturday strode brilliantly to center stage, the glory of the warm, dazzling Cape, captivating.

"Wow! It looks like it's gonna get crowded."

"Gonna? We should have gotten here sooner. It's gonna be hell trying to find a place to lie out, let alone save a place for…."

"Not to worry. They probably won't get out of bed, anyway."

"You're probably right. Um, looks like a perfect day to play in the surf. Wanta join me for a dip?"

"No, I'm gonna find a place for us to land."

"Okay, then. I'll be back in after I give the great Atlantic a try. Looks like there's a place over there near the big red umbrella."

When he returned, Trish was all stretched out on her back.

Holy shit. What a body! Her breasts aren't very big, but she has nice long slender legs. She really does justice to her bikini. I doubt she has an ounce of fat anywhere. Unbelievable! Sure would be nice to cuddle up with her, sometime.

"Back so soon? How was the water?"

"Fantastic! It's a lot warmer than the water off the Oregon coast. Ah, the sand, it's so...."

"Coarse?"

She squinted, rearing up to glance about.

"Um, it looks like the entire tourist population of Cape Cod decided to come to the beach today. Don't just stand there, come on and join me."

"Sounds good. I could use a little nap."

"I was just getting started before you got back. Ah, I'm still a little hung over from all the beer we drank last night."

"Me, too. I'd kinda like to get rested up for tonight."

"You'll need it. I hear the place we're going to is pretty lively. Diedra says they have a great band there on the weekends."

"Lively is good."

I wish I knew what I was going to do when I go before the Board of Officers next week. Spending four years in the Navy just so I can get a commission doesn't have the same appeal it used to. On the other hand, spending more time with her really turns me on. She's a real beauty, not like Judith. Her qualities don't shout at you, so much more discrete. Yeah, she catches your eye without stealing your soul, a lot like a magnificent painting or a great sculpture. She's just like Diana, outspoken and independent. Oh, oh, here we go again, another challenge. She probably has moves few people even understand.

"Are you awake?"

"Yeah. I was just resting my eyes, and, ah, thinking of you."

"How nice. Um, you said you weren't seeing anyone, but you kinda made it sound like what's her name, the one you talked about in your sleep, is pretty important to you."

"I haven't made any commitments."

"Nice try, but you managed to avoid my question."

"I've gone out with my share of women, probably too many."

His eyes darted to the left.

"Right now, officially, there's nobody."

"You say what's her name and you have stopped writing. Does it bother you?"

"Her name is Sandy. Um, maybe a little. Hell, I don't know. We've never really gone with each other."

"A minute ago, you said, you've dated too many. Ah, care to share?"

"Maybe I've overemphasized quantity instead of quality."

"Ah ha! So you are looking for the right somebody."

"And you'd like for me to believe you're not?"

"Touché! Ah, do you believe in love at first sight?"

"Sure, don't you?"

"Are you in love with me?"

His cheeks flushed.

"Not only are you curious as hell, but you cut right to the chase. Ah...."

"I'm not expecting an answer, yet. I just wanted to see how you'd react."

Rolling over onto her tummy to allow the afternoon sun one last shot at her back, she laid her head on the towel.

Under her breath, she mumbled, "You already have, you just don't know it yet."

"What did you say?"

Snickering, she looked at her watch.

"I said, I doubt if you know it, but it's time to go." Scrambling to her feet, she prodded, "Come on, I'll bet Diedra and Brent are waiting."

"Yeah. I'm getting a little hungry, anyway. What about you?"

"Famished."

Looking down on him, she smiled.

"We got off to a pretty rough start, but I want you to know, ah, I really like you."

"The feeling's mutual."

"Always before my father has had to put his stamp of approval on all of my friends and each and every guy I dated. Not this time."

"Is there a hidden meaning somewhere?"

"No, unless not caring what he thinks any more is a hidden meaning. Come on, we have to get back to our room, take a shower, and get ready for dinner."

"You say they have a good band?"

"So I've heard. You like to dance?"

"Very much."

"So do I. Well then, let's get going."

I wonder where all of this is headed. Trish's a great gal, but I didn't expect anything to come out of this weekend other than a few laughs and, well, I'm just looking to have a good time. Damn, I didn't expect to meet someone I really liked, ah, like a lot. She keeps asking about Sandy. How in the hell am I supposed to explain my relationship with her when I don't even understand it myself?

Outside their room, Trish toyed with the key.

"Brent's going to roll back. I guess he'll start all over again, from the beginning."

"Yeah, he told me on the drive up. Ah, I applied to roll back."

Frowning, she probed, "Oh! So, does that means you plan to roll back, too?"

"I don't know what I'm going to do."

He took a deep breath, looking directly into her eyes.

"I just don't know."

Taking his hand, she confirmed, "So, tonight is decision night, right?"

"Yeah, I guess it is."

"Just so you know, whatever you decide, you'll have my support. Here you do the honors."

She stepped back while he started to unlock the door.

"I guess I understand why you want to get your commission, but...."

He grinned, nodding his head knowingly.

"What in the world would ever cause somebody to commit to a four year cruise around the world on some stupid ship, huh?"

"It's more than a donation."

Turning the key in the lock, he opened the door.

"Spending time with you isn't making the decision any easier."

She stepped forward and kissed him lightly on the cheek.

Overwhelmed by their closeness, he turned and pulled her close.

"I've wanted to do this from the first moment I saw you."

When their lips parted, she looked at him for a moment and then quickly slipped by him, and made a beeline for her bed. Nervously she started going through her luggage.

"I've thought about it, too."

"I'm sorry, I guess I shouldn't have been so forward."

"Sorry? Why would you be sorry? I kissed you first."

She grabbed the bed for support.

"God, I'm shaking like a leaf. I'm feeling a little vulnerable, but if you want to know the truth, I'm glad you kissed me. I've wanted you to kiss me all day." Blushing, she confessed, "I like the way you kiss."

"You kiss nice too. Um, having you for a roommate is, ah, I really like the idea."

"Me, too. Um, I get first crack at the bathroom." Grinning, she suggested, "You'll have to move faster if you ever want to beat me."

"You probably need a head start."

"Only because I have more things to do to get ready."

"Anyway, while you're in there making yourself more beautiful than you already are, I'm gonna lie back and relax. By the way, how dressy is this affair tonight?"

A moment later she peeked her head around the door.

"What did you say?"

"I asked you what I was supposed to wear."

"Until a minute ago, I would said, 'Wear what I have on'."

Grinning, he teased, "Casual, huh?"

Later when they entered the nightclub, she winked at him as the hostess approached to seat them.

"Honey, I'm so excited you picked our favorite place to celebrate our anniversary."

He put his arm around her, planting a kiss on her cheek.

"Me, too. Brings back memories, doesn't it?"

"Celebrating number uno, huh?"

Trish snuggled close to him.

"If the next forty-nine are as good as the first, you'll be seeing a lot of us around here."

After they had polished off their steak dinners, Lee rubbed his tummy and groaned, "I'm stuffed. Don't anyone even suggest dessert. Not bad for a first anniversary celebration, right, honey?"

"I wonder what the next anniversary will bring? Hey, let's hit the lounge and show those college kids we still have some moves."

A packed lounge stocked with an energetic college aged crowd greeted them. After several dances and too many rounds of drinks to keep track of, Lee and Trish took a few moments to assess the lounge's appeal.

"I think you under stated the band's ability. They're fantastic."

"They are good, ah, so good, in fact, I think everyone visiting the Cape for the weekend is here."

"You're not so bad, yourself, Trish."

"You mean I passed the test, Arthur?"

"With flying colors. When you get a couple of mixed drink under your belt, you really come alive."

"You didn't know I was a party animal, did you? Come on, I want to dance some more."

"I don't know whether it's the mixed drinks or all the dancing, but I'm starting to wear down."

"Ah, too bad, I was just getting warmed up. Want to sit it out for awhile?"

"Yeah, I think I've had my limit."

I'd like to tell Trish how I feel about her, but, ah, no, I have to figure out what I'm gonna do about OCS, first. I also need to find out what's going on with Sandy. Sure would be nice to know exactly where I stand with her. Suppose it's possible to be in love with two women at the same time? Sure it is. I was in love with Sandy all the time I was chasing after Diana and Geri. Ah, yeah, I'd best not forget about Marianne? Sandy and I haven't made any promises unless agreeing not start without her is.... Nah, even though I don't start anything without her, she didn't say anything about seeing other people. No way of telling what's gonna happen between Trish and me, anyway. For all I know, this will just turn out to be a one-time engagement.

"Is something wrong?"

"What?"

"I asked if anything's wrong. For the past hour or so you've been so quiet."

"Nothing's wrong. I've just been thinking about what I'm gonna do about rolling back. What time is it?"

She looked at her watch.

"A little after ten, why?"

"I was thinking about calling my dad to talk to him about the review board I have to face next week."

"So, what's keeping you?"

"I was also thinking about what you said this afternoon on the beach. Did you mean it?"

"Uh-huh."

"Could you translate what really liking someone means to you?"

Blushing, she offered, "When I asked you if you believed in love at first sight, I asked you if you were in love with me, remember?"

"You said I didn't have to answer the question."

"That was then, and now is now. I'd like to know."

"I think it's possible to fall in love the first time you meet someone."

"So do I. What would you say if I told you I think I'm falling in love with you."

"I'm hallucinating or the drinks have really done a number on me."

"Why? Is it really so hard for you to believe somebody could fall in love with you?"

"Ah, no, but, ah, why don't we postpone this until I find a phone?"

"There's a phone out in the parking lot."

He glanced towards the parking lot. Suddenly his face drained of its color.

"What's the matter, Lee? Suddenly you're so pale."

"Damn, I'm really starting to feel it."

"Want some help?"

"If you don't mind."

"Not at all. Are you in condition to talk to your dad on the phone?"

"I think so. Yeah, he'll be glad to hear from me."

"Okay, if you say so. Come on, it's this way partner. Hang on. I'll get you there, somehow. Jesus, you must weigh a ton."

After struggling through the parking lot, she leaned him up against the phone booth.

"We're here. I suppose you don't have any money."

Grinning sheepishly, he asked, "Got a dime?"

A moment later, a voice droned, "Operator. What city, please?"

"I want to call collect to Salem. Um, that would be the Salem in Oregon."

"What number, please?"

"It's 363-4818."

"Who may I say is calling, please."

"Tell them their son, Lee Grady, is calling."

Four distinct buzzes later, a familiar voice answered, "Hello?"

"Will you accept a collect call from Lee Grady?"

"Ah, yes we will, operator."

"Go ahead, please."

"Hello, Dad? This is Lee, how's everything?"

"Fine. Are you okay? You sound kinda funny?"

"I'm peachy, Dad, just a little drunk."

He paused, winking at Trish.

"I called to ask your advice about something."

Chuckling, his dad probed, "Really tied one on, huh? How can I help?"

"I'm not doing so well with my classes. Ah, I'm thinking about rolling back and starting all over. What'd ya think?"

"It's a decision you'll have to make, and live with, son. We'll support you no matter what you decide."

"Thanks, Dad. It means a lot to hear you say it even though I kinda figured.... Ah, I'm in Cape Cod. A friend asked me to spend the weekend with...."

"From the sound of things, you've been seeing all the sights."

"A few. Ah, I can't wait to tell you all about it."

"So, have I helped you any?"

"Yep. Um, I really hate the Navy, Dad."

He winked at Trish.

"By the way, I met the girl I'm going to marry, someday. Whew, I'm a little drunk right now."

"I figured you might have had a couple."

"I told you I met someone special, already, didn't I?"

"You did. Are you going to be okay, son?"

"Yep, I'll be fine. Would you like to meet her?"

"Sure. She's pretty special, huh?"

"Uh-huh. Ah, she's standing here, right now. I'll put her on."

Lee motioned for Trish to take the phone.

"It's my dad."

"God, Lee. What am I supposed to say?"

"How about, hello. He doesn't expect you to say much."

"Um, hello, Mr. Grady? This is Trish, Trish Knight. Not like an evening, but like a warrior. You know, Sir Gallahad?"

"Hmm, pleased to meet you, Trish Gallahad."

Chuckling, she acclaimed, "You're funny, just like Lee. Um, I'm a friend of Lee's."

"So he said. How long have you known each other?"

"Not long. Ah, we just met yesterday."

"Hmm. I was under the impression he'd known you for...."

"Ah, no, but I feel as though I've known him for a long time. Your son's a very special person."

"Well, thank you, Trish. His mother and I are kind of partial."

"You have a right to be. Well, I think I'll let you go. I hope to talk with you again, sometime, real soon. Um, here's Lee."

Lee took the receiver.

"I think if I play my cards right, I can talk her into taking me home. You're gonna really like her. By the way, tell Mom I love you guys."

He hung up, wiping away the tears escaping down his cheeks.

"I don't think my dad believed me."

"Which part, having too much to drink or meeting the girl you plan to marry?"

Stumbling out of the booth, he mumbled, "Both."

"Oh, I don't think he would have any trouble believing you've had one too many."

She put an arm around his waist to offer needed support, and then started steering him in the direction of the inn.

"Come-on, cowboy, the time has come for me to take you to bed."

In the room, Lee wavered unsteadily in front of the bed and then plopped down upon it. For a moment he just sat there staring into space.

"Damn, If the bed would hold still I might be able to get undressed."

"Would you like for me to help?"

"Nah, I can do it. I've been getting undressed by myself for years."

"Am I supposed to leave or applaud as you strip?"

Squinting through glazed eyes, he suggested, "Crowd support is always good."

A minute later he was breathing heavily, rapidly disappearing into a coma.

Well, Lee Grady, last night I had to set the rules. Now things are different. I'm starting to fall in love with you.

She slowly slipped out of her dress, debating whether she should put on her pajamas.

Hmm, what difference does it make? He's too far-gone to care.

She slipped around the bed, turned out the lights, and then eased into bed.

The sun exploded through the window of their room. Cradled in the cavity of her warm body, sleep slowly receding, he awoke with a start.

"What the…? Where did you come from?"

She kissed him on the cheek.

"And good morning to you, too. I've been here all night."

The exhilaration was immediate as her lips gently brushed his.

"Now, this is what I call getting up on the right side of the bed. Um, did I behave myself?"

She kissed him again on the cheek.

"You were a perfect gentleman. Um, you were mumbling in your sleep. Sounds like you exorcised Diana."

"Hmm, I didn't say anything else?"

"Nope, not a thing. You didn't even snore."

"Lucky for you. Hmm, I think I'll take a shower and shave."

"Hurry up. If you play your cards right, we might have minute or two before they come knocking on our door to get us to join them for breakfast."

He returned a few minutes later smelling of mint toothpaste and shaving lotion.

"It's all yours. It's a good thing you don't have to shave. It's a bit steamy in there."

A few minutes later she peered out the door.

"My underwear, ah, it's in my suitcase, or are you going to be okay if I come to bed with just a towel wrapped around me?"

"I'll be okay, will you?"

She nodded, slipping quickly into bed.

"Um, did you mean it?"

"Mean what?"

"You told your dad you met the girl you were going to marry."

"Really?"

"You sure did. Do you want to take it back?"

"Four years is a long time to wait, don't you think?"

"It could be two. Besides, who says we have to wait?"

"Didn't you forget about your father?"

"I'm through letting him make those kind of decisions."

She leaned forward and kissed him.

"How would you feel if I visited you in Newport?"

"Encouraged, but when?"

"Soon. As soon as I can work it out. Ah, do you remember me telling you I could fall in love with you?"

"Aren't you afraid everything is happening kinda fast?

"Terrified, but I'd be more afraid of not facing how I feel honestly. Besides, it's possible to orchestrate when and two people are supposed to fall in love with each other."

"It's a big step."

"Uh-huh. Are you afraid of getting burned?"

"Maybe."

"Not much chance of getting a hit if you don't get in the batter's box and take a swing."

"There's also nothing wrong with taking a few practice swings."

"I agree, but I want more than a few weekend flings."

Her face started to redden.

"Would it sound crazy if I told you I want to make love with you?"

"A little. I, ah, I'm in no rush. I think we need think this over before we jump into something we both might regret. I really like you, Trish, but I don't want to rush things and mess everything up."

"You surprise me. Most guys are always trying to score. Thank you. It's nice to know I mean more to you than a quick roll in the hay."

Slowly she bent towards him and kissed him softly.

"You're special. I feel very close to you."

"You are. Except for the towel, you couldn't get much closer. Darn, I don't want this weekend to end."

"Hopefully this is only the beginning."

A sharp knock on the door dramatically influenced further discussion.

"Hey, you two. It's time to get up."

Giggling, Trish shouted, "Thank God. Saved by the bell. Ahem, who's there?"

"It's Brent. Diedra and I are headed for the restaurant to get some breakfast. Is everything okay in there?"

"Everything's just fine."

"Well, hurry it up, Lee, the bus is getting ready to head out of here."

"Be right there. Go ahead, we'll catch up."

On Tuesday, Lee was just finishing a letter to Sandy as Bob and Jim approached his bunk.

"Liberty started thirty minutes ago." Smiling broadly, Bob sighed, "Kinda nice being able to hit the beach every night. Um, writing to the girl back home?"

"Yeah, I owe Sandy a letter. It's been awhile since I wrote her last."

"Everything going good with her?"

"I don't know. I haven't heard from her in quite some time."

"What about the gal you met at Cape Cod?"

"I don't know. I really don't know. She might just be…."

"Hmm, might give Sandy a run for her money, huh?"

"Maybe."

Impatient as always, Jim frowned.

"Why don't we work out your love life over a tall cool one? The big city of Newport is calling."

AN OLD SALT

Falling in love, sometime akin to being struck by lightening, often happens unexpectedly. Sometimes immediate, preparing for the experience is impossible. What to do? It is the real thing? Are you blindly rushing into something you may regret later?

Newport's commercial hub was overflowing with tourists as the search for a place to quench their thirst began in earnest.
"Newport's usually not this busy. I wonder what the hell's going on?"
Jim dodged two ladies intent upon beating the rush into an antique shop.
"Looks like this sleepy old port town has come to life."
"I hope you guys know where you're going. With my vast knowledge of Newport, I don't have a choice other than rely on your knowledge of this berg." Lee chuckled, pressing on, "Knowing you, I doubt you spent all your time studying while you were on the base this past weekend. Surely you did a little exploring of Newport's digs."
Intent on his mission, Bob casually announced, There's a bar on up the street. Um, while we're working our way through this maze, how about telling me a little more about the gal you met at the Cape?"
"Not much to report other than she could be the real deal."
"Meaning?"
"For starters, she's a real beauty, and, ah, her body's not bad, either. Yeah, I really like her, but it's a little early to know if…. I wouldn't exactly

call a couple of days an adequate period of time to make a valid assessment."

"So, are you planning to see her again?"

"She and Brent's fiancée are driving down Friday to spend the weekend."

"You're not cashing it in with Sandy, or are you?"

"No way! Um, at least I don't think so. Then again, who knows, Sandy hasn't returned any of my letters? Maybe she's lost interest."

"Or found someone else." Shaking his head, Bob quickly injected, "Anyway, the postal service must appreciate your business. Lately you've turned into quite the letter writer."

Lee smiled sheepishly.

"Ouch, I get the message! Anyway, who's had the time?"

"Obviously not you. Fuck, will you look at all those little shops? My mom would absolutely go crazy here."

"If she's like mine, she'd spend the whole day and still want to come back to see if she's missed anything."

Chuckling, Lee teased, "It'd take a day just to cover all the shopping territory available to her in the great state of Texas."

"Funny! Anyway, this place is flat out awesome. Looks like pictures I've seen of the Colonial American period. Yeah, cobblestone streets and all."

"Hey, I just remembered, the America's Cup races begin Saturday."

"Hmm, I think you're right, Lee. We sure picked a hell of a time to look for a pub."

"How much further?"

"See those three guys coming out of the alley?"

"The pub's in an alley? You've got to be kidding."

"Don't believe me, huh? Well, you just wait and see."

Bob stepped in front of one of the guys emerging from the alley.

"Pardon me, but this is the alley where Trina's is located, isn't it?" Shaking his head, he added, "Christ, a man could die of thirst before he was able to find a tall cool one in Newport today."

"We just came from there. If you're interested in the Cup races, Trina's is the place to go. The Australian's just got into town. Seems they've kinda laid claim to Trina's. Man, those guys from down under sure can hold their beer."

Bob smiled smugly, turning towards Lee.

"See?"

The activity in the back alley pub was hectic, with all discussion focused on the upcoming challenge, finding a path to the bar nearly impossible.

"Oops!"

An elderly gentleman on his way to the bar bumped into Lee.

"Excuse me. I guess I'm a little too eager to quench my thirst."

"No problem. Looks like this place has standing room only."

"If you don't mind sitting with an old sea dog, you're welcome to join me. I got a table over in the corner with my name on it."

"Thanks. If you have room for three more, we'll join you as soon as I can get up to the bar to order a pitcher."

The old salt sneered at all the excitement erupting around them.

"The significance of the races means little to these landlubbers." Shaking his head in disgust, he continued, "If you'll help me carry, I'll spring for the first pitcher. On second thought, I'd better make it two. By the way, my name is Peter Joentengs. I retired out of the merchant marines two months ago after putting in nearly forty years." Beaming proudly, he boasted, "I earned my card when I was only twenty."

"Pleased to meet you Peter, my name's Lee Grady. Ah, these are my friends Bob Cardon and Jim Leffords. You come here often?"

"Most every night, I expect. You fellows grab a pitcher and the glasses and follow me. I'll show you to my table before those scavengers steal it from us."

"Lead the way, Peter. Looks like we'd better hurry, though. The vultures are already circling."

After taking a seat and pouring from one of the pitchers, Peter raised his glass.

"Here's to Trina's, the only pub in Newport where a good seafaring man is truly appreciated." Frowning, he continued, "The only time some of these glory chasers come in here is when the cup challenge comes to town. Once every four years to get drunk with some sailor they don't even know. You lads stationed at the Navy base?"

Bob nodded, surveying the activity surrounding him.

"Until we receive our orders to boot camp. Um, we dropped out of OCS."

"Didn't like it, huh."

"Not much."

"Ah, well. There ain't no harm in taking the short road. In the long run, it'll probably save trouble for everybody. Um, getting back to the races, you fellas realize the United States has never lost a challenge?"

"Really?" As he started to pour another round, Lee inquired, "Are they always held in the United States?"

"The winner always hosts the races. Yeah, you might say they're always held here. It's quite an event. A true test of seamanship steeped in tradition dating way back to 1851."

Peter could spin quite a tale. Holding his own with the best of those gathered for the pre-race festivities. Also, there was seemingly no limit to his ability to repeatedly raise his beer filled mug. Only a last call issued by the owners of Trina's managed to bring him to his knees.

"Well mates, here's to ya. May your ship always sail on calm seas and your heading nay be off course." Lifting his glass, he continued, "We may regret the last call a coming, but just round the corner is another glorious night oh raising our glasses on high again."

The last round didn't come soon enough, Wednesday's beautiful sunrise an unwelcome visitor. Bleary eyed and with throbbing temples, Lee reported to work. True to John's word, tomorrow was another day, as day passed into afternoon and prepared to blend into night. At the end of the workday, Lee was just getting ready for short visit to his bunk when Bob and Jim approached. Eyeing them as he plopped onto his rack, he frowned, rubbing his temples.

"I feel awful. Two more nights like last night and Trish will be coming to Newport to attend my funeral."

"You're not gonna let a little hangover from last night stop you from meeting up with our friend, Peter, again, are ya?"

"No, I'll be there. I just need a short nap, first."

Chuckling, Jim teased, "You were right, Bob. Bearcats just can't hold their own when it comes to a little serious beer drinking."

"Find out anything more about your plans for Friday with, ah, what's her name?"

"Trish? Yeah, I'm supposed to meet her at the Lamplighter Motel, Friday after I get off work."

"By then, you'll have plenty of tales to share unless John runs suddenly runs dry. He must have a trunk full of 'em."

"No kidding. I don't know where he comes up with all the stories."

Jim frowned, leaning against Lee's bunk.

"I'd like to know where he finds the room for all the beer he drinks. I've never known anyone who can put it away like he does."

Friday at 3:30, Lee knocked on the door of room 18. A moment later he was greeted by a ravishing beauty wearing black slacks and a white top. Handing him a glass of beer, Trish smiled warmly.

"It's so good to see you. Hurry up and get in here so the party can officially begin."

He leaned forward and kissed her lightly on the lips.

"I've missed you, too."

Diedra giggled, teasing, "Okay, you love birds break it up. You can put your bag over there on the chair. Come join us on your bed." Chuckling, she added, This time, I didn't have to twist Trish's arm."

"Yeah, hurry up. You have some catching up to do." Frowning, Brent inquired, "Is it true what I hear about you dropping out of OCS?"

"You've heard correctly."

"Are you okay with it?"

"Yeah, but I think I'll take it easy on the beer. The last couple of evenings have been a little hectic."

He crossed the room to put his bag on the chair.

"Getting out in two years instead of four more than compensates, Brent. Um, this evening after we finish dinner, I'd like to show you guys a place I stumbled onto this week. The Cup races are a real hot topic in there."

Diedra's eyes sparkled.

"Oh, neat. I love the Cup races. What's so special about your hangout other than the chatter about the Cup races?"

"It's where Australia's crew hangs out. Other than that, I guess you could say it's kinda salty in there."

"Are we going to be able to see the races tomorrow?"

"I'd like nothing better, Diedra, but unless you know someone who will give us ride out to where the race is held, we're out of luck. Looks like our only options are to hang around the docks tomorrow afternoon to hear about the outcome, or can catch all the pre-race activity tonight."

"Where are your recommending we eat, honey? I'm starved."

"Some friends of mine speak highly of Henry's Steak House."

"Sounds good to me. After another couple of rounds, we should all just about be ready for dinner."

During dinner at the steak house, it was a battle to get a word in edge wise. Finally Diedra captured the floor.

"I really respect your decision to drop out of OCS, Lee. The more I think about it, the more I like the sound of a two-year obligation."

Frowning, Brent probed, "I always thought you wanted me to become an officer."

"No, Brent, my father wants you to become an officer. Shortening your stay in the Navy wouldn't hurt my feelings one iota."

"I wish I'd known before I agreed to roll back."

"You guys can hash all this over later. Right now, I'm wondering when we're gonna visit Lee's favorite pub."

"Whenever you're ready to go, I'm game."

"Lead the way. You say you spent most of the week there?"

"Yeah, and I'm paying the price. By the way, you are gonna absolutely love the guy I met at the pub. Um, he's a regular there."

"What's so special about him?"

"His knowledge about the America's Cup is amazing. Yeah, and Peter's also very colorful. I think you're gonna enjoy meeting him, not to mention listening to his tales."

"Count me in. Just promise we won't stay all night."

"Why not? You have something else in mind?"

Blushing, she purred, "Maybe. Where's this place we're going?"

"It's in an alley about five or six blocks from here."

Diedra grinned, her eyes reflecting sarcasm.

"At least when the drunks get thrown out they don't have far to go."

Trina's was packed. Boisterous patrons, caught up in discussions about the upcoming races had turned the atmosphere into an animated zoo.

"There he is, over there. Peter's holding court in his usual booth."

The old salt rose from his booth as they approached.

"After all the time I've spent educating you about the sport of sailing, if you and your friends don't join me, I'll be deeply offended. They'll be no charge for listening to a history lesson about the America's Cup." Grinning, he casually added, "Course, I wouldn't turn down a mug of beer."

Lee studied the table that Peter occupied.

"Sure there's enough room for all of us?"

"Certainly! After a few adjustments, but first why don't you do the honors."

"Peter, this is Brent. The brunette beside him is Diedra and this pretty blonde beside me is Trish. Everybody, this is Peter Joentengs."

"Honey, how in the world did you find this place? I absolutely love it here."

"Ah, I guess I just kinda stumbled onto it."

"Trish, is it? Ah, yes, this place is nothing more than a dive for old sea dogs like me to drown their troubles. Course, every four years during the challenge races it livens up a bit. Correct me if I'm wrong Lee, but you didn't just stumble onto this place the other evening, now did ya?"

"No, it came with a strong recommendation, and ever since, you've kept me plied with beer and stories. I figured my friends would appreciate the environment in here as much as I do."

"Well, mates, from the looks of you, I can tell you have a thirst."

He raised his hand and motioned towards the bar.

"As soon as Mae gets a moment, she'll bring us a tall pitcher of draft with some more glasses."

Quickly Brent interrupted, "Why don't you make it two tall pitchers? Ah, and let us put it all on my tab."

After the waitress had made a delivery to their table, Peter started to describe how, from an early age he had acquired a love for the sea and a deep understanding of her strange moods. As he continued to ramble on, his knowledge about the America's Cup Challenge absolutely mesmerized all seated at the table.

"You see, the cup is all about tradition. For as long as I can remember, the New York Yacht Club has represented the United States in the competition, providing the best in sailing equipment and ships. The Club has instituted excellence in seamanship and yachting tactics."

"Come on, Peter, the New York Yacht Club represents our country because they have the most money."

Peter snorted with contempt, correcting Brent's misunderstanding.

"Not true! They win because they understand the meaning and the customs behind the history of the cup. They also win because they are willing to devote time and energy to represent our country. They have loyal members who are not afraid to dip deeply into their pockets to keep the tradition alive. It also helps that they manage to attract the best crew. Um, providing them with the best equipment helps, too."

"I agree with Peter, Brent. Daddy's a member of the New York Yacht Club. The way he tells it, some of those old farts are from families who have been members as long as there has been a challenge race."

"Your dad might be stretching the truth a bit, but there will come a time when someone from another part of this country will take the challenge away from them, and then, well, the cup races will no longer hold the importance it holds today."

Frowning, Trish inquired, "I don't understand, Peter, why?"

"Because they won't represent convention. They'll assume the Cup is for sale to the highest bidder and they'll look for ways to win even if it means breaking century old rules."

"Well, sounds like I've only scratched the surface in terms of learning about the Cup races. Ah, you can count on seeing me around here this next week, Peter. I might even buy you a beer."

Trish looked at her watch.

"Not like you haven't already spent enough time down here. Do you realize it's already midnight? It's already past my bedtime."

"Right. I guess it's time to go. As always, Peter, tonight has been interesting. You've been a source of enlightenment."

At the motel, Brent paused outside room 18.

"Do you want us to give you a call in the morning for breakfast?"

"As long as it isn't too early."

"What's a good time?"

"Nine should be okay."

He looked at Trish for confirmation.

"Um, on second thought, we'll call you."

"If we haven't heard from you by nine, you can expect a wake-up call. See you in the morning."

Trish closed the door, efficiently turning the dead bolt and securing the safety chain.

"If you'll turn on the lamp beside the beds, I'll douse the overhead lights." He grinned slyly.

"Um, which bed? Do I have a choice?"

"No silly." Chuckling as she moved quickly towards the bathroom, she teased, "Slow poke. You have to be quicker than snail to beat me to the bathroom. Are you gonna shower?"

"No, I showered before I left the base. I'm just gonna brush my teeth."

"You go first, then. I'm want to take a shower."

Moments later, she exited the bathroom, a cloud of steam trailing behind.

"Should I leave the lights on or off?"

"Why don't you leave the bathroom light on and crack the door a little?"

"Hmm, into peep shows, are ya. Hmm, I see you're not wearing an undershirt. I suppose you lost the shorts, too?"

"As a matter of fact."

"Uh-huh. Thought you might get lucky, huh?"

"No, actually as much as I've thought about a moment like this, I'm thinking we should wait. We have a lot of things to discuss before we start playing house. Um, I'll turn out the lamp as soon as you're in bed."

"Why leave the light in the bathroom on, then?"

"So we can talk, ah, unless you prefer the dark."

"Hmm, so, you think we should take a little time to get to know each other better, is that it?"

"Don't you?"

"Um, I was just debating which sleeping arrangements I prefer."

"Well?"

"I kinda liked the arrangements we had our last night at the Cape."

After closing the bathroom door so only a crack of light could escape, she turned back towards the bed. He had raised the covers to welcome her.

"Hmm, I don't remember this part of the play. Last time you were completely out of it by now."

"You could play fair and lose the towel."

"You think?"

Slowly she released the towel and started moving towards the bed.

Admiring her shapely silhouette, he gasped, "Damn, you're beautiful."

"Maybe sleeping in the same bed isn't the best way to slow things down, you know, talk about all the things you think we have to discuss, but…."

Slowly easing into bed, she confirmed, "I kinda like being close to you."

"Me, too. Ah, why don't you turn onto your side so I can cuddle you?"

After turning onto her side, she started inching towards the cavity of his warm, receptive body. Gently, he welcomed her into his arms.

"When I was a little kid, this is the way my mom and dad used to hold me when I'd come to their bed to get warmed up or to recover from a nightmare."

Lightly touching her shoulder, he coaxed her to continue closing the distance between them.

"Ah, so much better."

As he cradled her in the cavity of his warm body, he eased one arm under her head, letting it rest across her chest. Finally wrapping his free arm around her body below her breasts, he gently began snuggling her.

"Better? Are you comfortable?"

She snuggled closer, seeking closer union with his warm body.

"This is heaven. You're going to spoil me."

"All part of the plan."

Cautiously deliberating, he started to survey the silky skin of her back and shoulders with his nose, stalling intermittently to kiss her on the back.

"Isn't this where we left off?"

"Yeah, but I was in charge." Chuckling, she added, "After you passed out, I had my way with you?"

"Sorry I missed it. Um, so tell me about your father. I think he might turn out to be our biggest stumbling block."

"How right you are."

Pausing, she shifted uneasily.

"Um, he's very controlling. Ah, yes. Let me see. Where to begin?"

"How about the beginning?"

"Why not? Um, well, years ago my grandfather started an import-export business in Boston. After years of hard work, he turned the business into a very successful venture. By the time Daddy was in college, my grandfather had amassed a fortune. After my father graduated, he went into the business with grandpa and helped make the business grow even more. Now the business holdings stretch from coast to coast with some rather large holdings outside of the United States. Um, five years ago, when grandpa died, his estate passed on to his living heirs. Daddy received two-thirds of the estate, and, as the only living grandchild, I received the other third. As you might expect, Daddy is very protective of what he and my grandfather built."

"Your dad was an only child?"

"The only living one. All of his brothers and sisters passed before he graduated from college. Maybe it's the reason, since birth, I've lived a protected, sheltered life. In some ways, I've lived a life of privilege, attending private schools and finishing schools, but in so many other ways, my life's been very lonely. Yeah, maybe it's why I tend to be a bit rebellious, um, why I went to Brown when Daddy wanted me to go to Harvard."

"Are you sure I'm not just another rebellious moment in your life?"

"Not even. You're so much more. I told you I'm falling in love with you, and I meant it. Um, you know when you told your dad you had met the girl you wanted to marry some day?"

"Yeah."

"You haven't changed you mind, have you?"

"No, but I'm still in no hurry."

"That little admission would please Daddy to no end. He's not really excited about anyone, men in particular, getting close."

"Is he afraid somebody will hurt you?"

"Who knows."

"Well, I have no intention of hurting you."

"He's constantly reminding me not to associate with people below my station in life."

"Really! Well, I seriously doubt your father wouldn't like me, then."

"What's new? He has never, ever liked any of my friends. Except for Diedra, nobody seems to be good enough for me."

"Why did you decide to let me get close?"

"At the Cape when you called your dad, you were so drunk and so vulnerable. You told him you had met the gal you planned to marry some day.

I was so impressed with how you were willing share everything with him. You have very loving and trusting relationship with your family. By the way, I'm going to do everything I can to make your comment come true."

"Your dad won't like it."

"It doesn't matter any more. I'm tired of him running my life. I've made up my mind I'm going to be with the person I love. Yeah, aren't you lucky?"

"Yeah, but aren't you afraid of jumping into this too quickly?"

"There's no prescribed time limit for deciding who you want to spend the rest of your life with or when you've fallen in love, is there?"

He gently rubbed her back and shoulders with his nose.

"I suppose not. I just don't want to move too quickly."

Turning in his arms to face him, her eyes met his steady gaze.

"I'd suggest we rerun the movie, but last weekend we didn't get past the review of coming attractions."

She kissed him gently on the cheek and then on his responsive lips. After the brief exchange, she again fixed her gaze on his.

"It's okay if you want to make love to me."

"There's nothing I'd like more, but this might be a good time to reconsider before we go any further."

"No need. I've already decided what I want."

"You sure? There's no turning back once we've begun."

Pressing her body to his, anticipation causing her breaths to come more rapidly, she gasped, "I'm sure."

Gently placing his hand on her soft rump, he pressed his urgency against her erotic barrier as they began to passionately make oral union. The response was immediate and unguarded. Softly with long silky fingers, she began to knead his back and caress his neck as he continued to kiss her passionately, gently caressing her breasts.

"Oh, God! You make me feel so good. I want to feel you inside me."

The quiet environment was shattered with frantic urgency. In the blackness of room eighteen, two lovers responded to each other's touch and emotional clues. The tempo of the song they played rose and fell to a rhythmic beat, every contraction of their hearts, every unified breath producing a melodic harmony so pure in tone, so perfect in pitch. Like a good rendition of "The Flight of the Bumblebee," the notes were played with rapidity, the path to the end rising and falling with perfect accented clarity. A whimper and a loud rush of air escaped into the room. Just as it began, an eerie silence captured the dimly lighted chamber, concealing a tangled heap of clinging flesh. The models lie in exhausted repose, the frantic fury of their endeavor spent.

She stirred first, slowly seeking out the invader. With no incursion to quell, she petitioned for unity. Failed negotiations forced her to seek an extension to the conference, a momentary postponement.

"Hmm, it died."

"I sort of ran out of steam. Are you okay?"

"Oh, yes. It was wonderful."

Placing her arms around him, she hugged him tightly.

"I love you."

"I love you, too."

"Now you tell me. Sure it's not just because of what we did?"

"I'm sure. I might have fallen in love with you when I first met you."

An uneasy sensation invading his gut suddenly dampened his enthusiasm.

"Are we in trouble?"

"If you mean, could I get pregnant, I don't know. Could be. It's close, but I'm just not sure. Um, this is when we find out for sure if we love each other, isn't it?"

"I'd prefer for everything to be okay."

"But?"

"Is it enough for now for you to know I love you?"

"I suppose, for now."

She reached for him, expelling a rush of air.

"Hmm, feels like you're about ready to play some more. I love it when it wakes up. "

The big hand on the face of the clock neared six, the smaller hand seemingly stalled between eight and nine as rays of sunlight peeked through the small openings around the drawn drapes. Outside, a dog barked and a motel maid left the motel office headed for another vacant room to clean as a line of cars slowly made their way in the direction of the pier. Muffled sounds of chatty enthusiasts passing in front of the motel in pursuit of another day's shopping invaded the tomb-like setting.

Trish awoke first. Staring at the ceiling, she lay motionless, careful not to move lest she wake the inert form lying cuddled beside her.

Slowly she bent towards his ear and whispered, "Morning, sleepyhead."

Yawning, Lee stretched trying to shake off the mask of sleep.

"What time is it?"

"Breakfast time, still love me?"

Smiling affectionately, he whispered, "Sure. Today, tomorrow, and always."

He leaned forward and kissed her breast.

Chuckling, he teased, "With a little luck that is."

Gasping, she whispered, "I want to have your babies. If the time comes a little sooner than later, I'm okay with it, how about you?"

He shuddered, suddenly losing interest in the treasure before him.

"Isn't it a little early to talk about, ah, come on, let's get up and go get some breakfast. I'm starved."

Her eyes danced with excitement.

"Me, too. After breakfast, there's a little tour I'd like for us to take."

"Where you want to go?"

"I want you to take me to see all the mansions."

"Hmm, great idea. I've been wanting to see them since I first got here." Jumping out of bed, he boasted, "I can't believe it. I'm actually going to beat you to the bathroom. While I'm showering, why don't you give Brent and Diedra a call? Tell them we'll meet them at the restaurant in about thirty minutes. Um, you can be ready by then, can't you?"

"Why don't we make it an hour? I want to shower, too."

Diedra waited for her change at the cashier's station, turning to Trish with an inquiring glance.

"So, what are you two up to, today? Brent and I are going to do some exploring in the shops and maybe catch a little of the action down on the docks."

"I'd like to borrow your car so we can drive around the Newport area to see the mansions."

"Hmm, you probably want to be alone, huh?"

"Not necessarily. Do you and Brent want to go?"

"Nah. You go ahead. We'll take the tour this afternoon after you get back."

She grinned wickedly.

"So tell me, Trish, who's gonna do the driving?"

"Real funny, Diedra. You know I don't drive."

"Hey, Lee, I hear you're touring the mansions. You did know you have to drive, didn't you." Again smiling wickedly, she probed, "Or is that an issue Trish just forgot to share with you?"

"Why?"

"Trish doesn't have a license."

"You're kidding, everybody our age has a license."

"I think I'm going to enjoy this. Do I tell him, or do you want to keep it a surprise?"

Trish glared at her friend.

"Butt out, Diedra. I'll tell him. When do you want us to bring the car back?"

"What do you think, Brent? We probably won't go to dinner before six. If we leave at two to see the mansions, will we have enough time?"

"Sure. Lee, why don't you and Trish meet us for lunch?"

"When?"

"How about one o'clock? Ah, knock on our door so we know you're back."

A moment later, Lee started the red and white Buick.

"Hmm, nice car. So, are you gonna tell me why you don't have a driver's license?"

"Never needed one. Our chauffeur takes me wherever I need to go."

"A chauffeur, wow! So, I guess your family's pretty well off?"

"If you look at the houses we're driving past, many of them are similar to the one I live in."

"Hmm. The first time I caught a glimpse of these mansions, I was totally awed. They probably don't impress you much at all, do they?"

"Not so. These homes are classics. They belong to families who have belonged to the upper class of our society for several generations. You're looking at tradition, homes of people representing old money. Um, what are your thoughts about people from different backgrounds meeting and falling in love? Think it can work?"

"I've never thought about it. From what you've said so far, I would guess you're from the upper class, huh?"

"Some might say so. My family's wealth goes back two generations."

"And you said they're in the import, export business, so what about Diedra? Brent says her dad has lots of money."

"True."

"But, ah, your family has a lot more?"

"Probably, but I'm not really interested in who has or hasn't the greatest amount of wealth. It's all relative, anyway."

"Hmm, why is Diedra so intent on paying for everything? You probably could buy and sell her several times over."

"She has a need to impress people. Don't get me wrong. I love her to death, but sometime her needs to let everyone know she has money gets on my nerves."

"I'm afraid I'm pretty ordinary. Most of the time, I don't have two dimes to rub together."

"You have something far more important."

"Really?"

"There's a lot of love in your family. Like I told you before, it came through loud and clear when you had the conversation with your dad our last night at the Cape."

She smiled, lovingly squeezing his arm.

"Money's nice, but it can't buy happiness."

"I suppose not."

He started to fiddle with the keys in the ignition.

"Diedra and I are coming to Newport again next weekend."

"Wow! This could turn out to be a very nice habit."

"Indeed it could." Smiling, she probed, "It's amazing, isn't it?"

"What?"

"I've never felt like this before. Have you?"

"Kinda scary, isn't it?"

"Not any more. Right now I'm feeling very comfortable with my feelings and what I've decided to do."

"Which is?"

"If what I'm about to say scares you, you'd better say so right now, because, otherwise, I'm coming after you. I have every intention of making the comment you made to your dad come true."

"Have you told Diedra?"

"I just told her I really had a good time last weekend. Ah, I also told her I really liked you. I'm sure she'll give me the third degree on the way home."

"Lucky you. So, what are your plans for a career? Are you going to join your father's firm?"

"I don't know for sure. When I went to Brown, I majored in marketing with an emphasis in textile goods. My real love is fashion and design. Someday, I hope to be a buyer for a large retail store such as Macy's. I'm also considering a career in fashion design."

"How does your father feel about your choice of careers?"

"Daddy wanted me to go to Harvard and major in business."

"All parents try to shape their kids decisions to a certain degree, but I'm sure they want what's best for them."

"I agree, but not all parents are so controlling."

"Come on, he did let you go to Europe with Brent and Diedra. What was it, two years you stayed over there?"

"I still believe the only reason he let me go was because he didn't like the guy I was dating at the time."

"It's an expensive way to control a person's life."

"It wasn't exactly free for me, either. I had to promise I would join the firm when I got back."

"So, you're working for the company now, right?"

"I'm working in the division of our company responsible for importing textile goods from abroad. As you might expect, he wants me to live somewhere on the East Coast near Boston."

"What do you want to do?"

She reached over and squeezed his hand.

"Good question. You weren't in the picture when I made the agreement."

"Your dad's going to have my hide."

"Maybe not. I kinda opened the door."

"How?"

"I told my parents I met the man I was going marry you some day."

"They must have been thrilled."

"It's hard to say. Daddy's Sicilian, old school Sicilian. He doesn't allow his feeling to be known right out. I won't know how he feels until he meets you. As the head of the family, he's used to always getting things his way, including giving his blessing to whomever I date and the people I choose for my friends."

"You said he approved of Diedra. What about Brent?"

"I'm not sure."

The guy you were going with before you went to Europe, did he meet him?"

"No. I'd have to really love someone to put them through the ordeal of meeting my father."

"I may be able to understand where he's coming from by wanting to give his blessing to the person you marry. After all, marriage is a big step. I won't comment on the rest."

"Yeah, but there's a difference between him giving his blessing and actually selecting a husband for me."

"True. Um, you say you love me, and I've told you I feel the same way, but talking about getting married after only knowing each other two weeks is pushing things a little, don't ya think?"

Silence enveloped the car.

"Um, there's no reason to wave a red flag in front of your dad."

"No, but it would be nice to think he wants me to marry someone who loves me and wants to make a happy home for me and our children. I want him to accept the man I select."

"Well, maybe if we give it a little time, he'll come around."

"In the final analysis, it doesn't really matter. When I'm serious about something, he might as well support me because, in the long run, I'm going to do it anyway."

Lee's brow furrowed.

"My mom and dad brought me into a different world than you live in. I can't imagine what it would be like to have the things you take for granted."

"Lee, will you please pull over?"

He met her steady gaze after bringing the car to a stop.

"Well?"

"You seem to be skirting the issue. What's up?"

"I love you, but I need a little more time."

He leaned forward and kissed her.

"I don't expect a commitment now, but I want you to know I'm serious when I tell you I want to marry you and have your children." Grinning devilishly, she proposed, "I'll give you until next weekend. We'd better get a move on or we'll be late for lunch."

On Thursday, after work, Lee prepared to again hit the beach.

"Going to Trina's is getting to be a habit. Don't get me wrong, I'm not complaining."

"Enjoy it because when we leave for boot camp on Monday, and it's going to be long while before we have this kind of freedom again."

"True. On the positive side, it's another step towards getting out."

"I'm liking our duty here just fine. Just between you, the fence post, and me, I wish we didn't have to leave. Hurry up, will ya? Jim's probably going nuts waiting for us."

"Where is he?"

"He's gonna meet us outside Trina's, if we ever get there."

"Newport could be a little crazy tonight. Those damn Australians are just looking for an excuse to let off some steam."

"Like they haven't already. Um, how's it going with Trish?"

"If you can believe it, she wants to get married."

"Oops. How do you feel?"

"I don't know. I still don't want to give up on Sandy. Everything's happening so damn fast."

"Another reason to be happy about going to boot camp, huh?"

"Who's kidding. Don't get me wrong. She's a terrific lady and I really love her."

"But?"

"I'm not ready to get married. Not yet, anyway. How can a person have a life when they're in the service?"

"Things usually work out for the best. Just don't jump into anything until you're absolutely sure."

When they met Jim, it was evident he was almost in panic mode.

"What the fuck took you so long. We can't get in this way. The entry is jammed. It's packed in there."

"Why don't we go around back?"

Lee grinned, heading for the main street so they could get to the back door of Trina's.

"Wouldn't they call the back of this place, the back, back alley?"

In the back of Trina's, a door suddenly burst open. A man stumbled out, weaving unsteadily as he banged into Jim.

"What the…? Ah, is this the back door to Trina's, or what?"

"It's a zoo in there, man! The beer's flowing like a river. Um, try the door over there, but even if you make it inside, I doubt you'll be able to find a path to the bar."

Pushing their way in after navigating through a cluttered storage area, a quick survey indicated the bar was to the right some twenty feet or more away. Lee chuckled as he dropped to his knees.

"Well, it's time to hit our knees. I think we can make it to the bar if you don't mind crawling."

"Holy shit, Lee. It's so crowded I can't see anything except legs. You lead the way, Jim and I'll follow."

"Everybody's singing and dancing to 'Dancing Matilda'. Watch out that someone doesn't step on you."

Bob shouted, "This way, Lee. I see a small opening. Yeah, I can even see the bar."

When he reached the bar, he sat back on his haunches.

"Ya suppose If we reach up, maybe someone will hand us a beer?"

"What you guys doing down there, mate?"

"Trying to get to the bar."

"Just a minute. I'll get you some glasses."

A moment later, full glasses in hand, they accepted a hand up.

"If my mom could only see me now."

"What do you suppose she'd say, Bob?"

"Oh, something like, I didn't know you were so hard up for a brew. Can you believe this fucking mob?"

Jim elbowed his way beside them at the bar.

"Hey, looks like the big Australian over there is challenging the house to a drinking contest."

The Aussie stood on the bar holding a full pitcher of beer and raising his hands to quiet the house.

"We may not have beat you Yanks at the sailing end of things, but I've got fifty says I can out drink anyone in the house."

"You're on."

Slowly from the back of the massed humanity, a huge American challenger came forth.

"Well, okay, then. It's time to Ante up lads. Where I'm from, winner takes all."

The bartender banged a large hammer loudly on the bar.

"I'm holding. No matter who you favor, show me your cash."

A moment later, the last contribution received, he nodded at the contestants.

"Okay mates, listen up. Here are the rules. You each have thirty seconds to finish a pitcher. When the time's up, I'll hand you another. The one consuming the most beer and remains standing is the winner."

He held up a wad of green bills.

"Like the Aussie said, winner takes all."

With the huge throng urging them on, the two huge contestants started inhaling the contents of their steins. Slowly tipping them upward, they guzzled the contents until the last drops disappeared down their throats. After finishing a second pitcher, the American waved off the third pitcher, staring at the crowd. The Aussie wiped his lips, holding up three fingers. Suddenly the blank stare on the American's face was replaced by a look of horror. Panicked he looked for a place to retreat. With no place to go, he slowly bent over and braced his bulk on his knees.

"Thar she blows."

With the crowning of the Aussie as the champion beer drinker of Newport, an immediate chant arose from the assembled throng.

"New name, new name."

Finally an old salt leaped up on the bar.

"Trina's no longer a fitting name for a pub of such distinction. I say we give her a new name."

"How about, The Royal Thames Pub?"

"Here's to the Australian's courage and sportsmanship. From now on this place will be known as The Royal Thames Pub."

Friday afternoon before leaving the base, Lee fingered the letter Sandy had written.

Well, what do ya know? Sandy finally got the appointment she's been working so hard to get. Investigator for the federal prosecutor has a nice ring to it. She says she's gonna be investigating a Boston Mafia family with connections in Oregon. Hmm, looks like she'll be working on the case for years to come. She wants us to get together down the road. Getting out early didn't disappoint her in the least.

NOW AND FOREVER

For a short period of time or for the rest of your life, acquaintances influence the choices you make and the paths the option encourages you to take. Nurturing friendships, meaningful relationships, whether immediate or evolving over time, their influence remains the same.

Lee approached the booth where Trish waited.
"Been waiting long?"
"Maybe thirty minutes. I just finished a sandwich, are you hungry?"
"Maybe later. I think I'll just have a beer. Want one?"
"Sounds good. Um, it looks like you and I are on our own tonight."
"Diedra and Brent have other plans?"
"It appears so. Diedra told me they have a lot to discuss this weekend."
"I missed you."
"I missed you, too. Do you have anything special planned for this evening?"
"Nothing other than hanging out with you."
"What happened to this place? The town's a mess and this pub has a new name. I thought we won."
"We did. The owners decided to honor the Australian sailing team by renaming the pub. As for the town, it got pretty rowdy last night."
"And no doubt you participated, right?"
"Unfortunately. I'm sure glad the cup races are over. I don't know if I

could stand many more nights like I've put in the past couple of weeks. Um, how's everything on the home front?"

"My parents want to meet you. I don't think they're too happy about their little girl falling in love, but I suppose they figured it was best to pull in their horns and meet the competition."

"I'd like to meet them, too."

"Um, I told Diedra about us."

"You didn't expect to keep her in the dark forever, did you?"

"I suppose not. Um, she thinks we should slow down a little. She's afraid we're moving a little too fast. How about you, how do you feel?"

"We could be pushing it a little, but I'm not uncomfortable with the direction everything seems to be headed."

"Hmm, so you would agree there's no time table for falling in love or no prescribed amount of time two people are supposed to go with each other before they decide to settle down?"

"I suppose. Look, if you want to know if I still love you, the answer is yes. I fell for you the first moment we met."

"And?"

"Come on. Let's get out of here. This booth is too restrictive."

An hour later, lying beside each other, Trish lazily ran her finger up and down his chest.

"I love you. Being with you like this could become a wonderful habit."

"Could be, but it won't."

"What do you mean?"

"I got my orders. I'm leaving for boot camp on Monday. Looks like this is our last time together until I get out of boot camp."

"Hold me? I suddenly feel the need to be held."

"It's not the end of the world. Boot camp only lasts a few weeks."

"A week, a month, no matter, it's all too long. Um, I've been thinking about what we talked about last weekend."

"You mean about getting married?"

"Do you think it would work?"

"The life of a Navy wife isn't very glamorous."

"I don't care. I just want to be with you."

"What if I get sea duty after I get out of boot camp?"

"We'll handle it."

She inched towards him, looking longingly into his eyes.

"Well, what do you think?"

"About getting married?"

"Uh-huh."

"What about your father?"

"Nothing's going to change how he feels."

"What about your position with the company?"

"I'll continue working in the Boston office while you're at boot camp. When we know your duty assignment, I'll transfer to an office near where you locate."

"What are your options?"

"The company has offices in Boston, New York, Milwaukee, Los Angeles, Portland, and Seattle. We also have some overseas branches."

"What if my assignment isn't close to any of those places?"

"I'll quit. Being with you is far more important than any old job. Besides, I'll still have my stock in the company. We can probably live off the dividends."

He toyed nervously with the locket hanging from her neck.

"I don't want to lose you, Trish, and, even though I may seem to be stalling, I want us to be together. I absolutely adore you."

"Well, silly, now you know how I feel. Remember, I said it first."

"I still don't think getting married right away is the right thing to do. I'm having trouble seeing how waiting could hurt anything."

"Would you feel better about living together after you're done with boot camp?"

"Your dad would disown you for sure."

"It wouldn't be forever. Eventually you would have to make an honest woman out of me."

He leaned forward and kissed her. As their lips parted, he sighed.

"I don't know whether I like the idea of just living together."

"Well, if you don't want to get married right away and, as you put it, just living together isn't what you want, what in the hell do you want?"

"I love you, Trish and someday I want to marry you, but I want to wait until it's right. We could get engaged, though."

"Yes."

He shook his head, a sly smile creasing his face.

"I don't have a ring."

"Not necessary. So, are you asking me to marry you?"

"You're pretty persuasive."

She kissed him.

"I thought the way you kept stalling, you were looking for a way out."

She met his lips eagerly and held him tightly. As their lips parted, she giggled, her eyes sparkling.

"Honey, I don't care about the ring. If a ring is a big deal to you, we can go shopping at the grocery store and pick up a box of cracker jacks."

"Ah ha, you saw *Breakfast at Tiffany's*."

Nodding, she nestled closer to him and reached for his hand.

"I didn't know two people could fall in love this quickly. Are we normal?"

"I sure hope so. Ah, I have a proposal."

"You already proposed."

"How would you like to come home with me after boot camp and meet my parents?"

"Great idea!"

She pursed her lip in deep thought.

"I could meet you at the end of boot camp. Maybe my parents would join me so they can meet you."

"Yeah. We could get officially engaged, then."

"Yeah, it should give you time to find the right box of cracker jacks by then. Um we might even consider getting married right then."

"Whoa. You're moving much too fast for me. Let's get used to the idea of being engaged, first."

He grinned, shaking his head.

"When we go home after boot camp, we're going to travel by train."

"And?"

"We can make love to each other in five states."

"You'd better have a map handy."

"Why?"

"I want to mark all the sites on the map."

"Beats putting beans in the jar, doesn't it?"

She poked him in the ribs playfully.

"Are you hungry, yet?"

"A little, why, are you?"

"Yeah, come on. Let's go eat."

On Sunday, at three o'clock, he held her in his arms beside Diedra's car.

"Don't forget to write. I'm already missing you."

"I will."

She looked through her teary eyes into his.

"I can't wait until your boot camp is over. I don't care where you're stationed, I just want to be with you."

She smiled faintly, shaking her head in disgust.

"I'll start working on my parents, but in the long run, it still comes down to what we want to do."

Diedra and Brent approached, obviously excited about something.

"We've got some news."

"It had better be good. Staying holed up all weekend kinda made us feel like the poor relatives."

"Well, Lee, it looks like I'm going to be joining you. I'm gonna turn in my resignation on Monday. Then, I suppose, all I have to do is wait for my orders and go through boot camp."

"When are you getting married?"

"As soon as we can work it out. We want both of you to stand up for us if it can be arranged."

"You can count on it. Who knows? If you're not in too big of a hurry, we may join you."

YOU ARE MY SUNSHINE

...You are my sunshine, my only sunshine
You make me happy when skies are gray
You'll never know dear
How much I love you....

Seated on the bus Lee's mind wandered as Chief Margindar outlined their travel plans.

"When we get to Providence, we'll catch a flight to the Great Lakes Training Center."

Yeah, and then we find out what the real Navy is all about. I wonder what makes a bunch of OCS dropouts important enough for the Navy to assign a chief to ride herd on 'em? Right, no problem is too big or too small. I'd also like some advice about what I'm supposed to do with Trish. I really love her, but I'll be damned if I want to get married before I get out of the service. I could be convinced to move in together, but her father would probably put out a contract on me. Sooner or later, I suppose I'd better write to Sandy and tell what's going on. I'm not so sure anything's gonna work out with us, anyway, but...Damn, Trish has really complicated things. Lee's attention snapped into focus.

"Sailors, I have some news. It may not sit too well with you, so brace yourselves. Um, if you've been reading the newspapers, you know there's an apparent missile build-up underway in Cuba. Our government has told the Russians and the Cubans the build-up must stop and the missiles removed.

Gentlemen, even though your destination is Great Lakes, if President Kennedy can't resolve the crisis peacefully, you'll be coming back here to get your commission. Ah, during the first week at the lakes, you're gonna go through processing again. You fellas will be living in a temporary barracks while you take your service aptitude tests, have your dental work done, and undergo some medical tests."

The chief grinned, pausing to let his announcement settle.

"Yep, you just might face every variety and size of needle known to man. Buck up, mates, the time will pass before you know it."

"Fuck! I thought I was taking the fast track out of this man's Navy."

Lee glanced at Jim sitting in stunned silence and then back at Bob.

"Well, boys, looks like our stay in the Navy might just have changed."

Bob nodded, concern etched on his face.

"I sure hope cool heads prevail. All we need is to get involved in a war with those Russian bastards."

"No kidding. It could turn out to be the big one, the war nobody wins."

A heavy cloak of sadness suddenly enveloped Lee, making him shiver as he clutched the small heart-shaped locket Trish had given him. For the longest time, he sat gazing at her picture. Suddenly he clutched the locket tighter, almost fearful to let it go.

Until learning of the missile build-up, Lee had only paid passing attention to politics, rarely engaging in political discussions, even during some of the late night debates he so enjoyed during his college days. Even so, he was convinced the nation's citizenry admired and respected President Kennedy. The dynamic president, enormously popular, had revitalized the spirit of the American people. An increased social awareness seemed to have gripped the nation as the majority seemed to favor his blueprint for rebuilding the country's spirit.

Trish's dad is going to have a lot more to say about our plans than she thinks. There's no way he'll let us move in together, and I'm not so sure I blame him. Living together before getting married is a lot like shopping for shoes. In this case, I know we fit, but I'm not quite ready to pay the clerk and take 'em home.

Nothing the chief said prepared Lee for what he was about to face at boot camp, five defining events.

The odor emitted from the chow line should have given him the first clue. Unfortunately, even the horrific smell arising from the mess hall didn't dissuade him from eating. Two days into the week, he came down with a severe case of the runs. When he should have stopped, he ran. Running wasn't the problem because once started, he found it nearly impossible stop. He ran everywhere! Even when he coughed, he ran. On the third day, the doctor looked for the magic stopper to his running problems. Magically, the best solution was to pretend to take nourishment, going hungry far more appealing than constantly occupying the throne.

The needle made an impression. Apparently the singular case doesn't define the code doctors live by, the plural form coming closer. Never before had he seen so many needles, so many different sizes. Before he'd finished serving up his butt for pincushion, he was certain he understood what a goring in the bullrings of Spain was like.

The ritual military trainers impose to instilling discipline requires everyone to run everywhere they go. Unlike the former type of running, the training version was intended to move the entire body towards a destination. Maybe it was the relationship he had formed with the throne or maybe he was just stubborn. Whatever the motivating factor was, he opted to walk. The decision was immediately challenged by one of the service weeks.

"Sailor, on this base, you run wherever you're headed."

"I stopped running while I was in the brig for killing a Navy captain."

Lee sneered imposingly at the youngster whose authority had inflated his feeling of power.

"Is it necessary for me to tell you what I did when people told me to run while I was in the brig?"

Lee's unit wasn't required to run after the incident.

The Columbus Day storm struck Oregon with hurricane force. Unleashing a devastating attack, it left a path of destruction throughout the Willamette Valley. Telephone communication delays reached hours. When he finally was able to get a call through to his home, even Addison's somber tone was cause for relief.

"We're fine and our house is fine, but some weren't so lucky. The destruction of property is unbelievable. You should see Salem. There are trees down everywhere. The capitol grounds are a mess. People have lost parts, some entire roofs from their homes. Even the tower of one of the campus buildings on the OCE campus was blown down."

"You're sure everything is okay?"

"We're fine, don't you worry. Um, tell Bob we talked with his parents. They're okay, too."

"How's their home?"

"Hardly a scratch. Other than a few shingles, both of us came out of this whole mess in good shape. How's boot camp treating you?"

"I've faced worse. Um, what would you think if I brought Trish home with me when I come home on leave?"

"Trish, ah, the gal you told me about when you called to ask my advice about dropping out of OCS. Getting serious, huh?"

"She'd like to get married, but I'm dragging my feet. I'd like to get out of the Navy first."

"No need to rush. Um, we'll prepare the spare bedroom for her. Tell her she's welcome. We'd love to meet her."

When processing ended, Lee's unit was moved to a sparkling new barracks removed from the rest of the housing units. Bob shook his head as he put his gear on the rack above Lee's.

"If you can explain to me why we seem to be getting preferential treatment, I'd listen."

"Maybe it's because we may have to return to OCS if Kennedy isn't able to stare Khrushchev down."

Jim chuckled as he busily tended to his locker.

"Looks like we'll be able to stay abreast of what's going on. This place even comes equipped with a television set."

On a nightly basis, sometimes more often, the OCS unit was glued to the news reports reporting the latest in the confrontation between the USA and USSR. President Kennedy and Premier Khrushchev proved to be able adversaries. Making moves and countermoves in a tactical battle of wits and will, their chess-like game of cat and mouse gave the entire world cause to hold its breath.

The time passed slowly, but the letters flowed back and forth between Lee and the beautiful graduate from Brown University. He was mystified how rapidly their relationship was progressing.

News from Trish caused a huge knot to form in the pit of his stomach, as butterflies raced around like little kids looking for eggs at an Easter egg hunt.

Shit, she's been throwing up constantly for the past few days and she's late. Not good! Looks like she's gonna make an appointment with a doctor to find out what the hell's going on. I guess I'd better write and let her know we may not have any other choice than to get married.

Two days later, he received another letter. Something about the letter gave him the chills.

Guess I should read this one in private. Good, nobody's in the TV lounge.

After retreating to the lounge, he opened the letter and started to read.

Well, the cat's out of the bag. The rabbit died, and my fucking doctor decided it was his moral responsibility to enlighten my parents about its demise. Not too surprising, the doctor is an old family friend. Some friend, huh? If the truth were to be known, he's probably on my father's payroll. Needless to say, my parents are really pissed. I'm not even going to tell you how you're faring in our discussions. If you really want to get married, I'm all for it. However, if you're not really sure, we can still do what we planned and just live together.

He looked up and shook his head.

Fuck! Why in the not get married? Living together isn't the solution.

He looked down at the letter and continued to read.

If we do decide to get married, we could always find someone to marry us while visiting your parents on your leave. However, I'm not saying we shouldn't just live together for awhile until you get settled. No point in hoping for any support from my parents. All my mom can do is cry, and Daddy's damned Sicilian background has completely taken over. He's working longer hours at the office, and I doubt he has said a civil word to me in over a week. By the way, how does it feel to know you're going to be a daddy?

Unexpected news greeted him ten days before the end of boot camp.
Well, I'll be damned! Trish got her driver's license. She says she's

gonna drive up to New York to visit Diedra. Guess she's gonna be gone from Monday through Thursday and then catch a flight to Milwaukee to meet me. Hmm, Brent made it official and is coming to boot camp in December. Looks like he and Diedra are getting married shortly after Thanksgiving. Fuck, Trish's dad won't even buy her train ticket, so she's gonna splurge and buy a sleeping berth in a Pullman car for the trip. She still wants to mark all the important spots on the map where we make love. I like how that girl thinks.

Friday, the day the OCS unit graduated from boot camp, Lee checked with Bob to see what assignment he received.

"I got sea duty. Guess I'll be reporting to a ship called the *USS Tanner*, a coastal geodetic survey ship sailing out of the Brooklyn Naval Yard in Brooklyn. Hmm, New York's not bad duty. It sure could be a lot worse. What about you?"

Grinning, Bob boasted, "I got shore duty at the base in Norfolk, Virginia."

"Lucky you."

Lee stuffed two letters he received at mail call into his pocket.

"I'll read these later. Right now, we have some celebrating to do."

"Celebrate getting sea duty?"

"It's perfect. Yeah, Trish and I have decided to move in together. My assignment will allow her to work out of her company's New York office."

"So, who are the letters from?"

He pulled the letters out of his pocket and scanned the return addresses.

"Hmm, T. Knight and T. Knight. The date on the postmark is smudged. Don't know when she sent them. Oh, well, I'll wait and let her read 'em to me when I meet her later this afternoon."

"You sure? One of the letters is special delivery."

"Nah. They can wait. She's just probably telling me how excited she is about the trip home."

He casually stuffed the letters back into his dress blue's pocket.

"Well, what are we waiting for? The party at the pub in Milwaukee is beckoning. Good thing it's close to the hotel where I'm staying. I can check in on the way. What time does the train pull out tomorrow?"

"Eight o'clock. I hope she doesn't wear you out tonight. I'd hate for you to miss your ride."

Later, en route to the hotel to check in, Lee checked his watch.

Good, I have plenty of time. Trish doesn't arrive until seven o'clock. Let me see. If I leave the pub at six, I'll have plenty of time to get all ready before she gets there.

He patted the pocket where he had stuffed the letters.

She's going to be so excited when she finds out I have been assigned to the Brooklyn Navy Yard. In the hotel, after leaving the party at the tavern, Lee opened the first letter.

Hmm, her dad is a real asshole. I doubt I'm gonna have much of a relationship with him after Trish and I get married. He isn't even willing to acknowledge his daughter's pregnant. Maybe they're hoping it will go away. Like Trish says, old ways die slowly. Damned Sicilian. What's this? He's talking about sending her to the Seattle office?

The train sped westward towards Portland. Sitting alone in the semi-darkness of the domed car, a hunched figure fingered the fine linen texture of the stationery held limply in his left hand. The paper's fine blue hue displayed Trish's neat penmanship. Tears streamed down his cheeks as he reached into his lap to pick up the other letter.

His voice broke as he quietly read aloud its contents to confirm he had read it correctly the first time.

It is with great sorrow for your loss and ours I inform you Trish's life, as she knew it, ended in a most tragic manner. The state highway patrolman in charge at the accident scene told us a truck crossed over the centerline of the freeway and struck her car head on.

There is no way her mother and I can share with you the depth of our grief. Trish, our beloved daughter of twenty-three years, is now embarking on a new journey. Although different, I'm sure the loss for you is profound. I have been told your short, three-month relationship, implied more than a simplistic, impulsive relationship. I recognize to some, a period of ninety days represents a lifetime, and in Trish's case, apparently it was. You were the sunshine in her life.

Although she was firm in her conviction she had found the love of her life, I wasn't particularly pleased about your relationship. My feelings about such matters run contrary to my religious and family heritage. I'm hopeful, however, you will receive God's blessing.

Oddly, my daughter signaled a sense of something to come when she recently spoke of her final wishes. She clearly stated she wanted a simple ceremony attended only by her mother and myself. She wanted you to always remember her as being full of life. Her mother and I will expect you to honor her wishes.
Sincerely, Trush Knight

Lee gently folded the letter, placing it in its envelope. Slowly he ran his fingers across its fine texture, his mind wandering back to Friday afternoon.

Man, Mr. Knight must be an imposing son of a bitch. When he answered the phone, his voice has such a powerful tone to it. He spoke with such authority. I'd still like to know what all the shouting and that blood-curdling scream was all about.

Lee shrugged as he continued to review the conversation he had with Mr. Knight.

He was a lot nicer on the phone than I expected. I guess I can understand all of the commotion. Things must have been pretty emotional and confusing at their place. It isn't every day that a family loses a daughter.

His lower lip started to tremble.

Yeah, it isn't every day you lose the girl you plan to marry, either. Damn, I sure am gonna miss her.

Darkness engulfed the horizon as he began to drift boundlessly, without restraint. Only an occasional movement from the train affected his journey, an assessment of the enormity of love. The train's distant whistle faintly reminded him he was going home.

WINDOWS

A first love, not always passionate or influencing, introduces subsequent encounters. A first breakup evokes pain, not terribly different from later tragic losses to death or divorce. In either, torment enters the sensitive reaches of the human form, soon eating away at all resistance to anguish to expose vulnerability. Quelling the feelings of distress will require time and understanding, caring friends. Later, another person to love appears. It might replace the first, seemingly irreplaceable devotions, but compensation is never complete. A locked vault containing memories will always remain somewhere within.

Bob Cardon shook Lee from a deep slumber.
"Hey, wake up! We just pulled into The Dalles. There are two guys outside on the platform wanting to talk to you. One of them says he's your dad."
Jerking to a dreamy state of alertness, Lee stretched and yawned.
"Oh, God. I'm so tired. Um, what time is it, anyway?"
"A little after five."
Rubbing his eyes, he got up and stumbled towards the exit. At the exit, he held up his hand to shield his eyes from the blinding lights. He scoured the foggy boarding platform for his dad, his blurry eyes and the bright station lights making the task unusually difficult.
"We're over here, Lee. Wait a minute, we'll be right there."

He nodded, tentatively starting down the stairs.

"Those damn lights sure are bright. How are ya, anyway? I wasn't expecting to see you here."

"Good, I'm good. Ah, after we got your call, I didn't figure you'd be up for elk hunting. I suspect you might want some time to yourself. You okay?"

"Yeah, as good as can be expected."

"I couldn't resist the opportunity to see you before Wally and I take off up the gorge towards La Grande. Um, you've changed, son."

He smiled and then turned towards Wally, extending his hand.

"You didn't have to bring all this cold weather with you, did ya? It's freezing out here!"

Exhaled breaths rose skyward forming a thin cloud-like veil as they stood there enjoying the moment.

"So, you guys are gonna chase after the elusive elk."

Suddenly his shirt and trousers, his only protection against the elements, proved inadequate cover. Shivering, his teeth started to chatter.

"Damn it's cold. I'm not dressed for this crap."

"It's a little fresh. When I heard your train was pulling into Portland at seven, I figured we might as well head out early, try to make connections with you in The Dalles, and then make a dash for La Grande so we can get camp set up in time for the afternoon hunt."

"I wish I was going with you, but I probably wouldn't have my mind on it. Besides, a ten-day leave doesn't give me much time to visit with Mom and some other people I want to see."

Tears started to well up in his eyes as he half-heartedly rubbed his arms shield them from the gusting, icy wind.

"Yeah, I guess I'll just go home, get some sleep, and pester Mom a bit. Right now, I don't know what else I'll do." Lowering his eyes toward the wooden platform, he choked back a tear, "I loved her, I really loved her, dad."

Uncomfortable with the scene unfolding, Addison cleared his throat.

"I know son. Just remember, sometimes in this old life you have to make more than a two-bit wager. Occasionally you need to put six-bits on the line."

"Yeah, I know. Looks like I'm learning how to handle the six-bit wager. Yep, I really let it all hang out this time."

He looked away to wipe away a tear trickling down his cheek.

"Oh, I almost forgot to introduce you to my friend. Bob's the guy from Portland I've been writing you about." Nodding towards Bob, he raised his voice, "This is Wally and my dad." Turning back toward his dad and Wally, he suggested, "I'd like for you to meet my friend, Bob Cardon. Ah, he's the one standing over there under cover. Pioneers are always running for cover."

Bob, who had retreated to the entry to the train car, politely waved.

"I hope you'll excuse me if I don't come down to shake your hand, but it's colder than a witch's, you know what, out there. Ah, smart people like Pioneers don't make a habit of standing out in the cold with nothing on."

Addison chuckled, welcoming the humor.

"Pleased to meet you, Bob. When you fellas head back to your duty assignment maybe we'll have more time to talk."

"Sounds good. We're planning to ride the train back to our duty stations the Friday after Thanksgiving. I'd like to introduce you to my parents. By the way, I appreciate you checking up on them during the Columbus Day Storm."

"Glad to do it, Bob."

Addison cleared his throat, smiled and then thumped Wally on the shoulder.

"Well Wally, we had best be moving out cause those four-legged cows are awaiting, and the miles aren't getting any shorter. You take care, son. I'll be back the day before Thanksgiving, hopefully sooner."

He turned towards where their vehicle was parked, hesitated, and then turned back.

"It won't give us much time for a visit, but we can still celebrate your birthday and catch up on the past few months. By the way, why don't you give the attorney friend of yours, the one who works for the DA, a call."

He winked, conveying a silent but deafening message.

"She might just be what you need right now. I hear she has won a couple of real big cases. I guess she's a real hot item in the legal circles."

"I was thinking about it."

"You still interested?"

"I haven't had much time to think about it."

Stepping forward, Lee shook Wally's hand and then turned to his dad, unashamedly throwing his arms around him.

"Trish taught me how."

"I love you son."

"I love you, too, Dad. Good luck hunting."

Back in their seats of the train, Bob frowned.

"What in the hell was all the two-bit and six-bit stuff?"

"Hmm, it's hard to put into words. I guess the best way to put it is to say, ah, sometimes the good things in life cost a little bit more."

"I heard you dad mention something about Sandy. You gonna see her while you're home?"

"I've been thinking about it. God, it's been a while since I wrote to her last. She probably thinks I've lost interest."

"You haven't, have you?"

"Look, for your information, I really loved Trish. We were gonna get married after I got settled into my new duty assignment."

"Uh-huh. So, what are you gonna to do about Sandy? Trish's gone. It isn't like you were with her very long, anyway."

"I know, but it's just, well, somehow it just doesn't seem right."

Shaking his head, he persisted, "Do you still love Sandy?"

"Well, yeah."

"Damn it, Lee, we all think we're in love from time to time, but more often than not it's probably only because we get lucky and get a little. Even if it was as you say with Trish, I'm not convinced there isn't more than one person in this old life for all of us."

"You may be right. I don't know how many times I've thought I've met the right person only to discover when the lights went on everything was different. With her, though, it was different."

He stared into the distance, swiping at a tear starting to trickle down his cheek.

"She was so sure she wanted to spend the rest of her life with me. Um, she was pregnant."

"Damn! Ah, so it was the real deal! Damn!"

"Yeah, but her folks weren't too pleased. No, the truth is, they were totally opposed to us being together. If she'd lived, I would have been fighting the family bullshit forever."

He shook his head, contempt written all over his face.

"Her dad was one scary son of a bitch. I honestly think if I'd married her, he would have put out a contract on me."

The westbound train pulled into Portland's Union Station. Shortly after Lee had introduced Bob to his mother and brother, Bob's parents came scurrying towards them.

"Sorry we're late. There was a traffic accident on the Bandfield freeway. What a mess!"

Bob repeated the introductions and then turned to Lee.

"Well, Willamette, give me a call. If I have to ride one of these rail monsters all the way to the East Coast I might as well do it with someone I know, even if he is a Bearcat."

"I'll try to remember. You could help by writing your phone number down so I can remember it, though." Winking, he teased, "I never have been too good with figures."

With the tennis match knotted at love, they shook hands, deferring to another day.

"By the way, Bob, thanks for listening."

Lee hoisted his duffel bag on his shoulder, following his mother and brother to the car. Five months in the service had made quite an impact upon his appearance and demeanor. Now sporting one hundred sixty pounds on a sculpted frame, newfound confidence had added a spring to his step. An erect posture and military presence made him seem taller than his 69 inches. Tura nodded approvingly.

"I sent a boy off to the service about four months ago, and now he's returned a man."

"Well nothing stays the same forever, Mom."

"No? You still keep you hair cut short."

"True, but I still look like Clark Kent."

"Your glasses are more of an accessory, now."

In the back seat of their 1962 Chevy Impala, he settled back and closed his eyes as his mother began to navigate through the downtown Portland traffic. She looked in the rearview mirror.

"Before you go to sleep, I'd like to know what your plans are while you're home. Do you have anything special planned for this evening? Did your dad manage to catch up with you?"

He smiled, recalling previous conversations.

"You haven't lost your touch, have you? Yeah, Dad and Wally met the train in the Dalles, and I'll probably try to have a beer or two with Mel after I get some sleep." Yawning, he added, "I also have a couple of calls to make."

"Are you going to bed when you get home or would you like for me to cook up some breakfast for you?"

"Breakfast sounds good. Afterwards, I'm hitting the sack. The train ride wore me out."

"Didn't you sleep?"

"Bob and I didn't want to spend the money for a Pullman car, so we slept in our seats. Yawning again, he inquired, "How's school going, Greg?"

"Good! Mom's going to drop me off when we get back into town."

"Playing hooky to welcome me home, are ya?"

"Nah, I'll only miss my first and second period classes. How long are you going to be home?"

"My ten-day leave began when I left the base at Great Lakes. I suppose I'll have to leave the day after Thanksgiving if I'm going to report on November 27th at the Brooklyn Naval Yard. Do you suppose I can find a way to remember what civilian life used to be like?"

"I doubt you'll have any trouble."

Fifteen minutes later they merged into the outside lane of the moderately new I-5 freeway off the Capital highway exit. Usually an eye opening spectacle for him, this time the freeway passage turned into a nondescript forty-five minute snooze.

When his mother pulled into the driveway of their Marcia Drive home, he roused groggily.

"How long before breakfast?"

"I'll start a batch of French toast while you're putting your stuff away in your room, okay?"

"Sounds good. I haven't had French toast since I left home. Um, have you heard from Ron lately?"

"Not since he went back to school at OCE."

"Mom, we might as well clear the air. I'm gonna be okay. I'll miss her, but I know I'm going to get through it."

Tura stopped, turning to look at her son.

"We were looking forward to meeting her. It sounds like she was extra special."

"She was. Ah, I know this is going to sound strange, but somehow it all seems so unreal, kind of like waking up from a dream. Somehow I thought I would feel differently."

"You didn't know her very long and, ah, well, we all have our own way of grieving, Lee."

"I don't want to seem cold, but I know I have to get on with my life. I don't think she'd want me moping around forever, do you?"

"I'm sure you're right. You still have a lot of living to do. Did Addison talk to you about Sandy?"

"Yeah, he suggested I call her. Yes, I know, mother, I'll get around to it as soon as I get used to the idea of being home." Glancing out the window at the well-manicured yard, he asked, "Is Greg coming home after school, visiting friends or what?"

"He'll be home after wrestling practice and then probably go over to a friend's and work on some car they're trying to get up and running. I can't believe he's already a sophomore in high school. Seems like only yesterday that we were making a mad dash for the hospital. Oh, I almost forgot. Mel wants you to call. You have a few moments before breakfast is ready."

"He's a hell of a lot more thoughtful than I am. Would you believe I haven't written to him one time since I went into the service?"

He smiled sheepishly, shaking his head.

"In fact, we haven't talked or seen much of each other since we graduated from high school."

She smiled knowingly, waving at the phone.

"The French toast won't be ready for a moment. Go ahead and give him a call. He's working for his dad while he's waiting for his orders from the National Guard." Continuing to fuss over the French toast, she mumbled, "Seems like everyone is going in the service these days."

He reached for the phone, picked up the receiver, and dialed. The phone rang twice.

A familiar voice answered, "Hello?"

"Mrs. Marsters, do you have your strawberry crews set up for this summer yet? I'm looking for a job."

"For goodness sake, Lee, when did you get into town? When are we going to get to see you?"

"I just got home. Mom has some French toast on the grill for me, and then I'm going to hit the sack. I thought I would call Mel first, though."

"I'm glad you did. Oh, here comes Mel, now."

A short pause ensued, and then Mel's cheery voice invaded the silence, "Hi, there, you old salt! How's life in the armed forces?"

"Good, but I'm glad to be home. Do you have anything planed for this evening?"

"How does the Black Acre sound? I was thinking we could have a few brews, and catch up on old times. I need to find out how boot camp works because I report in two weeks."

"You want me to pick you up? I think I could talk Mom into letting me have the car for the evening if I agree to let give her a detailed agenda of our plans."

"Mothers are all the same. Tell her I'm going to kidnap you for the evening. I'll be by for you up about six-thirty." Chuckling, he teased, "I don't trust your driving."

Lee got up and walked over to the counter.

"Is the phone book still in the same place?"

"It's in the upper right hand drawer, Lee. I swear! You and your dad can never find anything."

"Not true. If you wouldn't move things around all the time, we wouldn't have so much trouble finding them. Hmm, here it is, the Marion County District Attorney's office."

"The French toast's ready."

"Would you put some butter on it for me? I'll only be a minute."

A moment later a monotone voice echoed through the receiver.

"District Attorney's office, how can I direct your call?"

"Could you please connect me with Sandy James?"

"Just a moment, please."

A moment later Sandy's cheery voice responded.

"This is Sandy James speaking."

"This is Lee Grady, Sandy. I hope I haven't caught you at a bad time."

"Oh, my, no, of course not! Oh, my God! How are you, I haven't heard from you for so long. I was starting to think you'd forgotten all about me."

"Too long. I'm sorry, Sandy, I've been a little busy."

"Tell me about it. You finally made it home on leave, I see. I hope you haven't made plans."

"I got in this morning. I thought I would give you a call to see if you still wanted to cook dinner for me like you mentioned."

"Look, I have a meeting tonight, but I have some time coming to me. I'll take the rest of the week off. Ah, would tomorrow work for you?"

"Tomorrow night would be great. I'm getting together with Mel tonight, anyway. I'm really looking forward to seeing you."

"I've been planning this moment for quite a while, Lee. Oh, it might help if I give you my address. Do you have you a pen and a piece of paper handy?"

"Fire away."

"Honey, you're not going to have a bit of trouble finding my place. I live at 629 Candalaria Drive, about three-quarters of a mile south of Mission Street. Just take a right off of Commercial. I live in a gray ranch style house with white trim. My goodness! I am so anxious to see you. We have tons and tons to catch up on. How does six-thirty sound?"

"Six-thirty is good. Um, I'd better let you go. We can catch up on everything tomorrow night."

At six-thirty, Lee rang the doorbell at the Candalaria address. A moment later, Sandy answered wearing an apron partially concealing an expensive looking green blouse and pair of tan slacks. As always, she looked fantastic. Her hair, normally worn down over her shoulders, was done up in a hairdo making her appear very professional.

Damn, she wears her beauty so much differently than Trish does, so much more mature and worldly. Before Trish hammered me into submission, she sort of reminded me of the girl next door. Sandy's more like a centerfold. She's already got me thinking about invading her space and we haven't even finished saying hello. Yep, she hadn't lost an ounce of appeal. She's still absolutely gorgeous!

She stepped aside.

"Get in here and let me look at you. My Goodness, it's only been five months and look at you. You were always handsome, but now, well, now you're a hunk! The extra pounds you've put make you look really good."

His cheeks immediately started to redden.

"Thanks. You're not so bad yourself."

"Same old Lee, when he starts feeling uneasy, those cheeks of his light up like a rose bush."

"I've thought about this moment for nearly six years."

She grinned, noticeably pleased.

"I'll bet you haven't exactly let any moss grow between your toes, besides, it's only been five or six months."

"Like you have, I mean, let moss grow between…?"

"I think it's best we leave our pasts in the closet. Um, well, don't just stand there. Take a seat at the bar. Would you like some wine? Ah, we still have a few minutes before dinner will be ready."

"No, I'm good."

He made a quick survey of her stylish, tastefully furnished home.

"Being an assistant DA must have its rewards. Your home is absolutely beautiful. Yes indeed, I'm impressed."

"Thanks. It's comfortable, it's mine, and it beats paying rent. So, how was your visit with Mel?"

"Good. We went to the Black Acres. For a black block building with a sawdust floor, it sure draws a crowd."

"Who's kidding. Did you eat dinner there?"

"I did. Um, I had one of their burgers."

"Yummy! You're starting to make me hungry."

Grinning, he teased, "Are you trying to tell me something about the dinner you've cooked for me?"

"Not really. It's just…. Ah, I absolutely love their flame-broiled burgers. Their steak isn't bad, either. So, I suppose you and Mel had a lot to catch up on?"

"Other than just an occasional get together, can you believe we haven't seen each other since our high school graduation? We talked some about going to graduate school together when I get out of the service. I think I'll get a certificate so I can teach."

"Fantastic! You would make a good teacher."

She eyed him nervously.

"Why haven't you written?"

"I've sort of been busy."

"It's okay to talk about it, Lee. Neither one of has been bound by any promises. It's just, well, ah, I've always hoped we would get together, someday."

"Me, too."

He stared at the stove in the kitchen for a moment and then dropped his head.

"I got involved with a really neat gal I met one weekend while I was visiting Cape Cod. She kinda swept me off my feet."

Her face drained of its color.

"I see. So, what are you saying?"

"Um, ah she died. Ah, she was killed in an automobile accident a few days ago."

"Oh, Lee, I'm so sorry. I didn't know. If I'd known, I wouldn't have brought it up. I'm so sorry."

"There was no way for you to know, Sandy. Ah, I'm kinda glad you did bring it up, though. I need to talk about it."

She smiled, nodding her head.

"I understand. Um, how about over dinner? I'm serving meat loaf, mashed potatoes, string beans, salad, and rolls. While I'm mashing the potatoes, I'll let you open a bottle of wine."

She reached for the bottle of wine and an opener.

"Here, make yourself useful. You can begin anytime you feel like you're ready."

"My pleasure." Grinning, teased, I'm a great wine opener. I get so much practice."

"Uh-huh. Um, I'm as hungry as a horse. I didn't eat lunch today."

"Why not?"

"I had to be in court for an appeal motion, and afterwards I had a plea and bail hearing. So, as usual, I did without, worked straight through lunch. Another typical day in the judicial system."

After Sandy finished mashing the potatoes, she motioned for him to join her in the kitchen.

"We're going to serve from the stove. There's not enough room on the bar for the serving dishes. Why don't you start by taking some potatoes? Don't be bashful, there's plenty. It doesn't taste near as good warmed up. Besides, my refrigerator is filled with leftovers. You'd think I would start taking my lunch to work so I could clean out the refrigerator before it all goes to waste."

"Why don't you?"

She chuckled, shrugging her shoulders helplessly.

"Who has the time? By the time I get ready for work each morning, the last thing entering my mind is putting up a lunch. Ah, getting back to, what's her name?

"Trish. You should know that she reminded me a lot of you. Um, I was in love with her."

"How long did you know her?"

"In terms of the time spent together, about as long as I've known you."

She nodded, staring at her fork.

"Long enough. Well, at least you kept your promise."

"My promise?"

"You didn't get started without me. In some ways, both of us have faced the termination of a relationship. Mine just wasn't as tragic."

"Care to share?"

"I don't really want to, but I suppose it's about time I got it all off my chest."

Smiling nervously, she confessed, "You're the only person I trust enough to tell. Wow! I never thought I'd ever admit it, but, ah, I just ended a rather messy involvement. Um, I'm not real proud to admit I got a little too close to someone from my office, a married man."

Tears started to well up in her eyes.

"We were working together on a murder case, spending a lot of time with each other. You know, night work."

"He got serious and wanted to get involved."

"Yeah, ah, he said he wanted to leave his wife and marry me."

"He probably told you he and his wife were through and there was no future for them."

"Exactly. Ah, I thought I loved him."

"So, what happened?"

"Before everything got completely out of hand, I told Eric I wouldn't be party to breaking up a marriage. Besides being unfair to his wife, can you imagine what it would have done to his children?"

"Children?"

Blushing, she acknowledged, "Yeah, he had three kids. Nice, huh?"

"I'm not gonna pass judgement. Um, how did he take it, you know, when you broke it off?"

"Not good. He told me he couldn't live without me. He didn't know how he could go on working in the same office knowing he couldn't be with me."

"So, what happened?"

"When he finally saw the handwriting on the wall, he resigned. Needless to say, it caused quite a bit of turmoil, caught everyone by surprise. He was on the fast track to the top."

"Did anyone suspect or know anything about your involvement?"

"Nobody that I'm aware of. Um, when Eric resigned, my boss told me he was promoting me to first chair, ah, lead prosecutor."

She frowned, again toying with her fork.

"What a way to advance a career, huh?"

"You didn't do anything wrong."

"Well, at least I didn't fuck my way to the top like some people I've known."

"Are you doing okay?"

"Yep!" Smiling, she confessed, "Now that you're here, everything is fine. Ah, I guess I told you I was asked to do some investigative work for the federal people in Portland. It could turn out to be huge!"

"I heard you were a hot item. Guess I heard right."

Settling into a state of muteness, Lee toyed with the string beans still remaining on his plate.

"I'm stuffed. The meal was great, but I can't eat another bite."

"I was afraid I had lost you."

"Trish and I were gonna live together. We had even talked about getting married, and, ah, she was pregnant."

"Wow! I don't know what to say. I, ah, I always hoped we'd get together someday."

"I wanted the same thing and then Trish came along. Before I knew it, we were…. God, things happened so fast. I'm not even sure…."

"Don't! It doesn't do any good to over analyze situations. What happened, happened. It's time to move on."

She hesitated, obviously detached and deep in thought.

"Ah, Lee, something's been bothering me. You say she was killed just before you came home on leave?"

"Yeah, the day I got my orders at boot camp, I got the letter from her parents."

"It just doesn't add up. You couldn't have had time to attend the funeral."

"I didn't. Ah, her parents, Mr. and Mrs. Knight had a private ceremony attended only by the family."

"Hmmmm. Odd, um, I suppose there was a reason?"

"In the letter, her dad indicated Trish wanted to be remembered the way she was when she was alive. She didn't like funerals."

"You don't say. Hmm, Knight, you say her last name was Knight?"

"Yeah, why?"

Sandy frowned. Suddenly she turned away and put her thumbs to her temples as she rubber her forehead with her fingers.

"And her father, what's his first name?"

"Trush." He signed the letter informing me of Trish's death, Trush Knight."

"And you say the Knights live in Boston?"

"I didn't, but, yeah, they live in Boston. Trish's dad is in the import-export business."

"Anatolia Enterprises!"

"What?"

"Nothing. I was just thinking out loud. Um, go on, I'm sorry I interrupted."

"Yeah, well, Trish told me her grandfather started the business, and when he died, the business passed on to Trish's dad. Now, I guess he has offices all over. They're located in LA, Portland, Seattle, and Milwaukee, ah, not to mention his offices on the East Coast."

"Or the holdings abroad."

"You act like you've heard of him?"

She shrugged casually, her eyes darting to the left.

"No. I was just thinking of someone that I have to meet with next week. You'll have to excuse me, but when I'm working on a case, I sometimes drift in and out."

"I really never ever got over you, Sandy, but, ah…."

"We didn't made a commitment, no promises. Besides, I sort of got involved, myself. Come on. Let's go into the living room. I'll clean up the dishes, later."

"You know, Sandy, life is really strange. I don't think I've figured out whether it's anything more than a bunch of stops and starts or if it's only a continuous episode without a real beginning or end. Maybe what we think is the end is nothing more than a disguised form of another beginning."

She grinned, nodding her approval.

"Deep. Tonight, you're much to deep for me."

"The day before I went into the service, ah, you were right in what you did."

Her cheeks reddened as she fixed her eyes on the floor.

"Why, because I stood you up?"

"Were you involved then?"

She reached out and took his hand.

"You don't have to sit so far away. I promise I won't bite."

"Wow! Our first promise. Um, so, why did you back out?"

"We shared a brief but wonderful experience in Eugene. I didn't want anything to tarnish the memory."

She puffed up her cheeks, exhaling.

"I guess I want something more, far more. This is a lousy time to bring it up, but, ah, do you think we a future?"

"We've been on this path for nearly six years. I'd kinda like to see where it goes."

"Okay, so, now I want to know how you feel about having something more. Right now, both of us are a little vulnerable and my timing stinks, but

I don't want to take another chance of losing you. Ah, I'd like to start thinking about spending a lifetime together."

"What are you trying to say?"

"I love you. I have since we first met. With my career's headed in the right direction and with you just about to finish up your tour of duty in the Navy, I'd like to start talking about something more than these five or six year reunions."

"I've always hoped we could be together."

"I've felt the same way. Waiting like we have hasn't been easy."

"In the long run it probably has been for the best."

"I'm not going to let you go this time, sailor. I'm not talking about making promises, but I'm ready to start talking about our future together."

"I get out of the service in another eighteen months."

"And I'm counting the days. Ah, the reason I don't want to make any promises is because I don't want to tie your hands to some commitment you might regret. You have the right to still do a little shopping."

"And you don't?"

She grinned, her cheeks suddenly turning a deep crimson.

"Do me a favor?"

"Anything."

"When the times right, knock on my door? If I answer, I'm telling you I want us to be together, forever."

"You've got it."

He took her hands and slowly allowed his eyes to meet her steady gaze.

"About your refrigerator. You were telling me about all the leftovers. It sort of reminds me of our personal lives."

"How so."

"We've never really finished anything. Both of us have a lot of…."

"You're right. I think it's about time we both got rid of the leftovers."

She slowly rose from the couch and took off her apron. Letting it drop, she reached for his hand.

"Come on. I'm want to show you the bedroom."

She smiled, leading the way down the hall.

"I'll need a moment to freshen up. I promise not to be long."

"I know the drill. I'm not supposed to start without you."

His leave passed too quickly, nightly visits with Sandy proving to be a catalyst for forging a future. When she agreed to join his family for Thanksgiving, the signs brightened immeasurably.

Addison smiled at the surprise dinner guest.

"When Lee was in college, a visit from the assistant DA would have signaled something a little more serious than simply breaking bread."

"Whose to say my visit isn't serious? I didn't come to inform him of his rights, but I'm not here to merely pass the time of day, either."

"Oops, your comment must have touched a sensitive spot. Lee's starting to light up like a Christmas tree. So, Miss James, I take it, this isn't just a professional call."

"Please, call me Sandy. Um, no, it's completely social."

"What does your future hold?"

She smiled coyly, carefully concealing a pat poker hand, "With Lee or in the DA's office?"

"You're good. I see why you win the cases you try. Will you be joining us when we take this guy to Portland this evening, Sandy?"

"Thank you for the offer, but I think I'll say my goodbyes when he takes me home from this fantastic dinner."

She toyed with her fork.

"Lee only has eighteen months before he gets out of the Navy. We both have things to sort out and, well, I hate goodbyes."

"The way you are advancing in the DA's office, I suppose you have a lot of demands on your time. Your career doesn't give you much time to consider forming a serious relationship, does it?"

Her cheeks reddening, she retorted, "I don't think I could ever let career aspirations get in the way of marrying someone I love."

"We would be premature to consider a relationship right now, though. Both of us have a bump in the road to overcome."

RELEASING FRUIT

Trees release their fruit to signal the beginning of another harvest while menacing clouds diffuse moisture to indicate another storm is at hand. Shedding tears, signifying grief, portrays the loss of a loved one and the human species, the most protective of all God's creatures, foretells a new generation with the emancipation of its children to face the world on their own. Letting go breaks your heart but failure to do so hardens the soul.

The train conductor hollered, "All aboard. All aboard for Boise, Cheyenne, Omaha, Chicago, and New York."

Lee shook his brother's hand.

"Be good. Give them hell in wrestling and baseball."

Turning to his mother, he gave her a hug.

"Take care Mom. Ah, remember not to cry this time, and Dad, next time I'm home, we're going elk hunting together."

"Don't forget to write to Sandy, son. She's a keeper."

"I'm sure Sandy and I will manage to exchange a few letters between now and the time I get out. I'll write as soon as I get settled. It's okay if you beat me to the punch, though."

The clatter of the train soon claimed another victim as the skyline of Portland gradually disappeared from view. A couple of mixed drinks in the train's lounge on top of too little sleep soon found Lee in the midst of a trip into dreamland.

The marvelous sight of a potato harvest unfolding, not unlike a successful Broadway production, was performed on a grand stage. Drawn curtains slowly opened, the rising sum announcing the beginning of Act One. Mother Nature's orchestra, a gentle autumn breeze carrying the blended sounds of crickets and birds, struck up a melodic introduction. The play, at last, had begun.

Huge machines moved slowly across the landscape leaving large brown deposits of freshly dug potatoes in their wake. Row upon row of those Russet Burbank delights gradually started to form as the tractors, a Deere, an old Case, and some other brand, unearthed the buried bounty from one end of the field to the other. Spuds, increasing mounds of them, steadily started to appear as far as the eye could see. The harvest tempo, in full swing, ground on at a seemingly never ending pace. Some of the crew had elected to cook for the harvesters, neighboring volunteers, while others opted to toil in the fields. None the less, all facets of the harvest, a continuum of related activities, merged into purposeful movements. When finally the digging concluded, the men and older children began filling burlap sacks with the freshly dug potatoes lying exposed in the field. Once the sacks were full, the laborers carefully set them upright in the rows so women following behind with huge needles and loops of heavy twine could start sewing the sacks shut.

Finally ready, carefully placed in neatly aligned rows, the bagged treasures stood erect awaiting the drivers of the flatbed trucks to start maneuvering through the harvest deposits to pick them up. Forced to stop every six to eight feet so workers on both sides of the truck could hoist the cargo to awaiting hands, trucks moved steadily onward. Repeated exertions from hot sweaty field workers strained against the bulk of each hundred-pound offering, time after time lifting sacks to those waiting so the bundle could be quickly added to the neatly designed crisscrossed puzzle forming on the truck beds. All the while, the late afternoon harvest scene displayed an impressive sight, trucks waiting to be loaded standing framed in the field where shadowy reminders of a setting sun had been cast.

When finally ready for transport, the drivers put the lading machines in gear. A groaning surge against the mass resting on their decks is drowned out by the whine of overextended motors as the burly movers slowly quit the fields. All through the late afternoon, plumbs of dust billowed skyward as the trucks repeatedly set a course for a distant storage cellar, a dirt-mounded structure resembling a large Arctic family dwelling. Similar to the icy castle's cover of snow, this grass covered shelter provides insulation against the extreme fall or winter temperatures. Inside the cool, dark crypt, smelling like clean dirt, is the temporary resting-place for the shipment. Done! The cellar door closed with a creaky report. The huge master lock clicked, giving a signaling another harvest lie sheltered and secure, waiting for a motion, a notification of a fair market price."Huh? Market value? What's Lisa's going rate?"

"Wake up, Lee. It's time for breakfast. Besides, neither you or I can afford her price."

Frowning, Lee confessed, "Hmm, I must have been dreaming."

"Yeah, just like I was when I decided to tackle Lisa."

A few moments later Lee and Bob sat down for breakfast.

"Well, Bob, did she wear you out last night?"

"Lisa wanted me to rent a sleeper compartment. Incredible as it may seem, I passed. She begged me to put the coal to her, but I sure as hell wasn't gonna pay just to ride her all the way to Cheyenne."

"What's in Cheyenne?"

"Her next gig. Seems, ah, Lisa from Seattle is a prostitute."

"Like you didn't know. It was painted all over her face."

"Yeah, anyway, I'm gonna scope out Norfolk for a little action. When I've got the lay of the land all figured out, I'm gonna invite you down for a visit. I guess you know if you want to find out where the action is, you have to put your trust in the hands of a Pioneer."

"Just like last night, huh?"

If Bob accomplishes only half of what he thinks he's capable of, he'll set the world on fire. He shrugged. *Oh, well, what can you expect from a pioneer? Guess I'll write a letter to Sandy. Everything is starting to look up, at least we seem to be moving in the right direction.*

Lee's second visit to the big apple lacked the auditory and visual affects he had previously experienced. He shook his head in disappointment as they entered the tunnel.

A passenger sitting next to him chuckled.

"If you're looking for a breath-taking introduction to this great city, never arrive by train. The best way in is by ship, then, it's a tossup between coming by car or airplane."

He frowned as he looked out the window.

"This is the poor man's porthole. All you see is railroad tracks and slums. However, if you want to experience something really awesome, step out onto the street for a few minutes after you've wandered around inside Grand Central Station."

"I doubt I'll wander far. I've got a duffel bag and a big suitcase to tote around."

"Check 'em in a locker. Trust me, you don't want to miss the view of the city coming out of Grand Central for the first time."

Without the burden of his baggage, Lee found movement within Grand Central Station's crowded expanse much easier. At the information booth he got directions to the Brooklyn subway line and the name of the stop closest to the Brooklyn Naval Yard. Finally, unable to wait any longer, he searched for an exit.

Well, I hope it's as impressive as the guy on the train said it would be.

Stepping outside onto the sidewalk, he immediately was caught up in the hustle and bustle of nonstop activity. Overwhelmed, he emitted a gasp of veneration, time seemingly stopping as he stood there staring at the activity about him. People moved like ants in every direction while a tangled web of private automobiles and taxis jammed the streets. Horns blared, tires squealed, and cars darted in every imaginable direction. All activity was seemingly intent on setting a private course somewhere. Awed by the activity of the big city as it again unfolded before him, a sudden urge to look upward overpowering him. Slowly his eyes moved upward, immediately re-focusing to accommodate the extended distance they had to survey. Like an express elevator, they leaped from the deep shadows cast by the tall skyscrapers to the top of the nearest structure. Momentarily squinting to allow his eyes time to acclimate to the bright reflection of the partially hidden sun, he studied the upper most structure, as, unconsciously, his feet began to move in choppy, circular steps so he could take in the entire skyline. His visual thirst finally quenched, his neck aching from gawking upward, a sheepish smile spread across his face.

Damn. Is this impressive or what? I'll bet chiropractors are kept busy here.

A cabby parked out front hollered out his window, "You going somewhere or are you going to stand there and stare at the window cleaners all day?"

Lee smiled as he turned to reenter the station.

"Just taking a look, but thanks anyway."

After reclaiming his baggage, he headed for the subway entrance to the Brooklyn line. Forty minutes later, he exited the subway station.

I'll be damned. There's the Watchtower Building. Don't suppose people need divine guidance to ride the subway in New York, do ya? Hum, like the lady at the information booth said, there's the Brooklyn Naval Yard down the street. The YMCA should be just around the corner and the Naval Authority Building up the street from there. Damn, this suitcase is heavy. I guess I'd best get rid of it before I report. I doubt I can have civilian clothing aboard ship, anyway. From now on, the only difference between Clark Kent and me is Clark changed in a telephone booth. I don't have to contend with the stupid cape, either. After renting a locker at the YMCA, he hoisted his duffel bag onto his shoulder and headed towards the Authority Building to report.

After logging Lee's time of arrival, the duty petty officer revealed, "Well, sailor, I have some good news and some bad news. The good news is you mustered in on time. The bad news is the *USS Tanner* is still out on maneuvers somewhere in the North Atlantic."

He checked his log again, looking up with a sly smile.

"Uh-huh, you have the honor of reporting aboard the *USS Argot* for temporary duty."

"How temporary?"

"Until the *Tanner* gets back into port."

"What do you know about the *Argot*?"

"The *Argot* crew is getting ready for the ship's annual inspection. From what I hear, things are pretty hectic. They're trying to do about a year's worth of maintenance in two weeks. You may not see the streets of Manhattan until the *Tanner* returns."

"How long before they get back?"

The petty officer again checked his book.

"Um, it appears they're due back mid-December. The Navy didn't do you any favors by assigning you to the *Tanner*. I'm told the *Tanner* has logged more sea miles the past two years than any ship in the Atlantic Fleet."

"We joined the Navy to see the world, didn't we?"

"I suspect you'll see more of the world than you ever bargained for."

"Thanks for the information."

The petty officer pointed in the direction of the YMCA.

"The front gate to the Navy yard is down the street to your right from the YMCA. The Marine guard at the gate will tell you how to get to your ship from there. Ah, always give those fucking JarHeads straight answers, no bullshit. Those bastards are mean, and they don't take shit off anybody. If you're inclined to fuck with people, pick on somebody else."

After reporting aboard the *USS Argot*, Lee was shown to his sleeping quarters.

"Pick out an empty rack and put your gear on it until you get around to storing your gear in one of those lockers. After dinner will be soon enough."

The seaman waited for Lee to locate a rack and hoist his duffel bag onto it.

"Follow me and I'll introduce you to the duty boatswain. Boatswain's mate Helms will tell you where to report in the morning. He'll also take you to the chow hall. They start serving any minute now."

"Looks like the Navy is as advertised."

"How so?"

"Everyone gets a rack and three squares a day."

"Don't forget traveling the Seven Seas and a gal in every port."

A moment later, Jones introduced Lee to Matt Helms.

"Pleased to meet you, sailor. Looks like you'll be working for me until your ship comes in. You give me a good day's work and we'll get along fine. Slack off and I'll have your ass for lunch."

"What do you want me to do for the rest of the day?"

"We've knocked off for the day. You can either go back to the sleeping quarters to stow your gear, or you can accompany me to the chow hall. If you're gonna go on liberty, I'd eat now."

"How's the food?"

"Damned good! Seaman, if you plan to hit the beach, I wouldn't overdo it. Working for me with a hangover isn't much fun. Here, grab a tray and follow me."

A few minutes later, Lee followed Helms to a table where two other seamen sat.

"Mind if we join you?"

The taller of the two looked up slowly, almost mechanically.

"Feel free. I'm Bill Pence and this is my friend, Auggie Milaski. Hmm, since Boats is giving you the tour, I guess we'll all be working together on the deck force."

"Grady just reported aboard. Like you, he's been assigned duty aboard the *Argot* until the *Tanner* gets into port. Bill and Auggie, this is Lee Grady."

"Pleased to meet you. Auggie and I are going down to Times Square for liberty tonight. You're welcome to join us if you like."

"When are you leaving."

"As soon as we get showered and change into our blues. Have you stowed your gear yet?"

"No. I just threw it up on my rack."

"You'll have plenty of time to stow it while we're showering. Where you from?"

"Salem, Oregon. I started out at OCS."

Auggie nodded, slowly placing his utensils on his tray.

"Ah, did seamanship get you, too?"

"Yeah. Ah, I don't remember seeing you at OCS?"

"We were in the other company."

Bill grinned, slowly picking up his hat.

"Boats had us chipping paint today. You'll love it. Can't remember when I had so much fun."

"Yeah and tomorrow I'll show you how to paint, Pence, the Navy way."

An hour later, after departing the ship, Lee straightened his tie as they headed towards the gate.

"Would you mind if I make a stop to make a telephone call before we head up the street to the subway?"

Auggie chuckled, his eyes twinkling.

"Gotta clear things with the little woman, huh."

"No, it's just a friend. I'll only take a minute."

Bill grinned, stooping forward to catch Lee's eye.

"Good looking?"

"Engaged! "She is a friend of a girl I used to date. I figured the least I could do was let her know I'm stationed here."

Lee dialed the phone number listed beside the name, A. James.

After three rings, a voice replied, "Hello? This is Diedra speaking."

"Diedra, this is Lee Grady."

There was a long pause, a deafening delay in the phone service.

"Hello?"

"My God! Ah, I was just thinking about you. It's so nice to hear your voice. Ah, how are you? What in the world are you doing in New York?"

"I'm fine! I just reported to my duty station at the Brooklyn Naval Yard. So, when's the big day?"

"Brooklyn, my you're close. Ah, Lee, Brent and I got married last week."

"Trish mentioned you were getting married before…. Ah, she told me it was going to be this week or the next."

"We changed our plans."

"Well, congratulations. Wow! Ah, so, how's Brent doing?"

"Brent's doing as well as can be expected. You know what boot camp's like. Ah, you didn't get an invitation because we didn't have a clue about how or where we could reach you."

"I was on leave. Besides, it sounds like you might have been in a hurry. Ah, who'd you get to be your maid of honor?"

The silence was deafening.

"Diedra, are you still there?"

"Yeah, I'm still here. Ah, I asked a sorority sister, Patti Anatolia. She and I went to Brown together."

"Hmm, her name sounds familiar. Was she in the same class with you and Trish?"

"Um, yeah. So, are you doing okay?"

"It's been tough, but I'm moving on. Ah, I guess you knew we were moving in together after I got out of boot camp?"

"Lee, you gotta know Trish was never the same after she met you the weekend we spent at the Cape. The girl was really in love."

"I loved her, too. What a way to go, ah, you know, the accident."

"Accident? Oh, you mean the accident."

"What did you think I meant?"

"I guess Trish told you how she and her parents fought about you."

"I knew they weren't exactly pleased. What does her parent's feeling about me have to do with the accident?"

"Oh, nothing. I was just thinking about how nasty everything got before she came to New York to visit me."

"She said they weren't too happy about me, but she never said anything about it going any further than them expressing some concern. Ah, did you know she was going home with me to visit my parents?"

"She mentioned it. She wasn't sure whether her parents would allow her to go."

"Hmm. It's funny she never said anything. The last I heard, she was still planning to come. She had already purchased a ticket. Well, anyway, I still feel bad about not attending her funeral. I guess you didn't go either?"

"Ah, no."

"I guess her dad told you she wanted a private family service?"

"That's it! That's exactly what her dad told me. Ah, you know Trish. She wasn't one for ceremony."

"Right. Ah, have you seen or talked to her folks lately?"

"No. Um they moved to Seattle the day after Brent and I got married. I guess Mr. Knight plans to run the business from there for awhile. You talk about doing things quickly."

"Hmm, I'm surprised Trish never said anything about Mr. Knight's plans. When did you find out about the move?"

"Like I said, after the wedding."

"Hmm. You'd think Trish would have mentioned something about his plans when she visited you in New York. I also don't know why she requested a private service. You'd think her family would have at least wanted you there. You were her best friend."

He paused. A long, deafening pause that seemed to consume an eternity.

"Are you okay?"

"Damn! I can't help but feel like I deserted her."

"Believe me, you didn't desert her, Lee."

"I suppose you're right. Um, her dad mentioned she seemed to sense something was about to happen to her."

"I know exactly what you mean. Come to think of it, I'm sure she knew something was about to happen."

"How so?"

"I don't know how to explain it, exactly, but she was really down in the dumps when we were together in New York."

"Did she say anything?"

"Well, when we were at Macy's shopping, we ended up in the infant's department. Ah, she looked at me with tears in her eyes and told me she would never be able to shop for baby clothing with you. She told me things were going to be very different. At the time, I didn't understand her comment, but now, it's perfectly clear."

"Hmm, I guess she didn't tell you."

"Tell me what?"

"Ah, you didn't know she was pregnant, did you?"

"My God, no." She hesitated. "So, that's the reason."

"What? The reason for what?"

"Ah, the, the reason she, she was sort of sad. She was a Catholic, you know. She came from a very traditional Italian family."

An uncomfortable silence grabbed the phone lines.

"Lee, even though it's sad about Trish, you're better off."

"I can't understand why you would say such a thing. I loved her."

"Her dad would never have let you two get married. Her accident, well, let's just say it might have saved you a lot of grief. Her dad can be real mean-spirited when things don't go his way."

"Shit, from what you're telling me and what Trish told me about him, he sounds like some sort of Mafia Don or something."

"Lee, I really have to go. It's been great talking to you."

"Yeah, I should go, too. Some guys from my ship are waiting for me to join them for some exploring in downtown Manhattan. Um, let's stay in touch."

In an off-Broadway bar, Lee studied his new companions.

Now there's a huge man. Funny how he always slouches whether standing or sitting. Everything he does is at half speed. I wonder if all people from Southern California are so laid back. If I didn't know better, I'd swear he could take a nap between strides. I guess that's called doing it at Pence speed. If Bill ever starts talking about something he really cares about, the story could last all night. Funny how much Bill and Auggie are alike. The only real difference is the amount of space Auggie occupies and the way he talks about his background. Oh, and his polish accent is priceless. I'll bet he's a hoot when he had a few. Yep, they both seem content to just meander through life.

Auggie Milaski, a jolly, robust native of Wisconsin, spoke proudly of hailing from a small Polish community just outside Milwaukee. On their maiden voyage to the bowels of Manhattan, his eyes twinkled when he spoke about his drinking bouts in Shaboinski's grove.

"Ya, I sure do enjoy drinking peeva with my friends on weekends at the grove. Sometime we even got lucky."

He smiled broadly, his eyes dreamily reflecting as though miles away. "I really enjoy good times and drinking beer, um, two of my favorite pastimes."

"Interchangeable acts, no doubt."

"Hey, you two. I have a proposition to make."

Auggie grinned. "Well, hurry up. I'd like to get to bed before morning, Bill."

"Um, if you'd like, we could go to the USO on Friday and pick up some vouchers for a hotel room. They also have free movie tickets and passes to plays there."

"Bill, don't forget the passes to see the Knicks play in the Garden."

"All you have to do is go in there and tell them you want free tickets. They post all of the attractions available for each day. There's one draw back. To get the tickets, you have to show up in your uniform. If they have tickets, they're yours."

"Sounds like the USO is a bargain basement of sorts."

"Auggie and I stayed in a hotel before we reported. Going through the USO we got the room for less than half of what you'd pay normally."

"Count me in. I've a lot of things I want to see and do while I'm here. A three way split will make it almost affordable."

"They might even have one of those get acquainted parties next weekend, don't ya think, Auggie?"

"Ya, the USO brings in girls to mix with the guys, and, I hear a few keepers even show up once in awhile. What are you interested in seeing while you're here, Lee?"

"I'd like to start of with the Statute of Liberty, the Empire State Building, and the United Nations. I've already seen Tiffany's, but visiting Wall Street or taking a ride on the ferry at night to see the skyline would be neat. What do ya think?"

"You can see the skyline from our ship."

"I doubt it's the same from the ship, Auggie."

Bill grinned as he continued his slow saunter down the street.

"Sounds like you're gonna be busy. So, I guess we're on for the weekend, huh?"

The routine on the *Argot* never varied, but hitting the beach to take advantage of the free offerings through the USO or sharing a few beers in a favorite bar made the constant banging on metal to remove paint or slopping on a coat of paint bearable. On weekends, Lee and his new comrades weren't forced to return to the ship. This freedom enabled them to study and learn the personality of the big city in ways not possible on their weekday liberty calls.

Friday, at liberty call, they headed for the USO in Times Square after finishing another uneventful day of trying to hide the invasion of rust. Ed Rowel, another shipmate headed for duty aboard the *Tanner*, joined them.

Surveying the offerings on the activity board at the USO, Lee frowned.

"Hmm, not much available tonight. Ah, Ed, you looking forward to the *Tanner's* return?"

"January can't come too soon for me. I'm really getting tired of chipping paint."

"I hear ya. Damn, I was looking forward to seeing the Knicks play. They must be out of town. Well, guys, what should we do, now?"

"Auggie and I are probably gonna take in a movie."

"Yeah, I see there's still a couple on the board we haven't seen."

Bill peered over his glasses.

"Would you and Ed like to join us?"

"Nah, I'll take care of the hotel, and then I'm gonna go visit the Empire State Building. I haven't been up to the top yet."

"I think I'll tag along, Ed. I'll save the United Nations for tomorrow."

"While you're out and about, why don't you check out a good place to get some Chinese food." Grinning, Bill added, "I've a real yen, get my drift?"

"Ya, look for something that's cheap! I don't give a damn where we go, as long as the place serves a lot of food. Just make sure it doesn't cost an arm and a leg."

After splitting up, Ed and Lee set out to get a room.

"I've been to the United Nations, but I've never taken the tour."

"Tomorrow will be the first time I seen it other than in a picture. I plan on taking my time. I don't want to be rushed, so if you plan to go with me Ed, don't be in a hurry."

"After we get the room, we can see what kind of Chinese joint we can find on our way to the Empire State building."

A short supply of cash was inspiration enough, but the snarl of traffic cast the deciding vote for using a common New Yorker's mode of transportation, Shank's Mare, to scope out the restaurant and navigate the remaining distance to the Empire State Building. When they finally reached the top of the Empire State Building, Ed scanned the horizon reverently.

"Man, New York sure looks a lot different up here, doesn't it?"

"No kidding. I never realized it before, but Manhattan's just an island surrounded by four other islands. It's so narrow!"

Grinning, Ed explained, "They call those islands, boroughs. What Manhattan lacks in width it more than makes up for in length."

He peered over the rail.

"Everything seems so small from up here. Look at all those people down there. They look like a bunch of ants. One of these days I'm gonna come up here at night. I'll bet the view is spectacular. Ah, maybe we should start heading back towards the hotel. By the time we get there, it'll almost be time to go eat."

Four hungry sailors entered a family owned off-Broadway restaurant called Chin's.

"The bellhop at the hotel told Lee and me that this place is his favorite. It's also supposed to be easy on the pocketbook."

"Sounds good. At a time like this, money's not an issue. I just want to get enough to eat. These joints never seem to serve enough to eat."

In the dimly lighted dining room, the Chinese waiter handed them the listing of the evening's fare and four glasses of water. While they scanned the neatly printed menus, he stood by patiently waiting for their decision.

"This place doesn't have those special meals I'm used to." Turning to the waiter, while pointing at the menu, Bill inquired, "Don't you have meal number one, meal number two, and so forth? I'm talking about the meals with three or four different kinds of Chinese food on one plate."

"No understand. We serve what on menu. No meal number one."

"I guess you don't have the other two, either. Hmm, I guess I'll have to order one each of my favorites."

Auggie smiled, nodding with appreciation.

"At these prices, even if you order three or four selections, you'll only have to shell out about six bucks."

"Waiter, I'll take an order of Pork Chow Mein, Pork Chop Suey, Pork Egg Foo Young, and Pork Fried Rice. Ah, why don't you give me a bowl of Won Ton Soup to warmed up on." Closing his menu, Bill probed, "So, what's it gonna be, Auggie?"

"I'll have the same. Does tea come with the order?"

The frail little Chinese nodded, his eyes twinkling as he finished recording Auggie's order.

"That be two orders of chow Mien, Chop Suey, Egg Foo Young, and Fried Rice. I bring four plates."

"Don't forget the orders of Won Ton Soup."

Bill looked at Lee and Ed.

"You guys make up your minds, yet?"

Ed chimed in, "Make it three."

"Ah, what the hell, why don't you make it four? I don't know how big the orders are, but the price sure is right."

The little Chinese fellow frowned.

"Two more? Too much!"

Jabbering in Chinese he shook his head in frustration.

"Too much! You order too much."

Turning helplessly toward the kitchen, he shrugged his shoulders, petitioning for help.

"They order too much!"

Dismissing the diminutive host with a wave of his hand, Bill countered,

"Maybe we order too much, and maybe we don't. No matter, why don't you stop jabbering and hustle into the kitchen and rustle up our orders? We're starved."

Bill's impressive height and sizable bulk, quashed the confused little Oriental's spirit to protest. His concern registered, he picked up the order book, and shaking his head, turned towards the kitchen. Mumbling inaudibly in his native tongue, he made a hurried retreat. Behind the swinging door to the kitchen, excited, humored voices, all seemingly jabbering in unison, accompanied the banging and rattling of pots and pans. Reverberating from within like a large orchestra warming up before a performance, the din of frantic activity was punctuated every so often with giggles and muffled expressions of disbelief.

"Too much, too much! They see. They order too much."

Moments later, the waiter returned with the orders of Won Ton Soup and some tea. For a moment, he stood there searching the faces of four sailors.

"Too much! You see! You order too much!"

Bill chuckled as the little waiter again retired to the kitchen.

"That poor little fellow has a lot to learn about this group. I don't know about you guys but I'm so hungry I could eat a horse."

"Ah, Bill, he's a little guy and probably doesn't eat much."

"Damn, this soup is good."

Lee glanced towards the kitchen as a cook appeared in the doorway with an empty cart. Their waiter followed closely behind pushing another cart filled with covered platters.

"Holy shit. He wasn't kidding. I think we order too much."

Glancing at the parade of food approaching, Bill shrugged.

"If you can't handle it, let me know."

The waiters cleared the table of everything except their silverware, laid out the plates, and then started putting the platters of food on the table.

"Too much. No room for all you order."

"Now there's a meal. Just put a platter of each entrée on the table. You're only got one job to do, little fella, you just keep it coming."

Nodding, the waiter obediently put a serving dish of each order on the table, retreating to patiently wait by the cart. As a serving dish emptied, he hurriedly replaced it with another, again retreating to stand by his cart. From the entry to the kitchen, the assembled staff murmured in amazement as platter after platter disappeared at an alarming rate, excitable jabbering in the incomprehensible dialect of their native tongue urging them on.

Moments later, Bill looked up from his partially full plate, and then he stared at the cart.

"Damn, we're only about half done, and I'm running out of room. I'm afraid you got the best of me. Too much."

"Ya, me, too."

Auggie's eyes twinkled.

"Could we get some boxes for the rest of it?" Patting his stomach, he confirmed, "Too much."

"Yeah, bring out the doggie bags. I'm stuffed."

Lee frowned, shaking his head.

"Think you can actually face this stuff again tomorrow? I'm so full, I doubt I'll be able to eat for a week."

The waiter smiled beckoning for the entire kitchen staff to join the formal ceding ceremony.

"Too much! See, I tell you. You order too much."

Bill grinned, accepting the check.

"We'll be back! Mark it on your calendar. We'll be back next month. Next time, we'll know better than to order so much." Moaning, he added, "I don't know about you guys, but I'm going back to the hotel and flop out on the floor. A little shut eye is starting to sound pretty good."

Back at the hotel, Lee reared up and looked at the clock. Sleepily mumbling, he turned over and fluffed up his pillow.

"Why doesn't someone turn off the TV? It almost midnight?"

"I'm gonna finish watching the news, and then I'll turn it off. Are Auggie and Ed asleep?"

"I'm just straightening out my bag. Auggie's in the bathroom."

An unexpected knock at the door interrupted the commercial break.

Bill lazily suggested, "Somebody answer the door. I'm only wearing shorts."

"I'll get it! You want a blanket, Bill, or you plan to entertain the caller in your skivvies."

As Ed tossed Bill a blanket and headed for the door, Lee rolled over in bed so he could catch a glimpse of the late night caller. Timing the move perfectly, he caught a glimpse of a tiny sprite of a figure framed in the partially open door.

Ed petitioned, "Ah, can I help you?"

Dressed flamboyantly, she posed seductively.

"All depends."

Impressively wearing the doorway, she scanned the room.

"Hmm, three. What do ya know? This is my lucky night. Well, are you just going to stand there and stare or are you gonna invite me in."

Broadcasting the arrogance and assurance of a seasoned late night veteran, she seductively took an uninvited step into the room.

"Who's first?"

Startled by the invasive forwardness, Bill rose from his station on the floor, wrapped the blanket around his waist, and slowly ambled towards the bathroom.

"First for what?"

The tiny little nymph sneered.

"What do ya think, big boy? Go on and get your pants on. When you're all dressed, you can leave so I can be alone with the guy on the bed. He looks to me like he's all primed and ready to go. Hurry up, I don't have all night. I'm on a break."

Auggie peered around the bathroom door as Bill stumbled in.

"What's going on?"

"Four? Even better! Hmm, I don't do group peep shows, so hurry up and clear out. You'll all get a turn."

"Where are we supposed to go?"

"How about down the hall, Mr. Doorman?"

Lee jumped out of bed and scrambled into his pants.

"I'll join you, Ed. Um, you can start with Bill or Auggie."

Bill slowly exited the bathroom and headed for the door.

"I'll let Auggie go first, he's dressed for the occasion."

A smirk spread across the hard but vulnerable face of the shapely sprite.

"I'll send your friend after one of you when we're done. Ah, since this is break time, old Gina's gonna give you all the special rate."

At the end of the hall, Bill grinned as he looked back over his shoulder.

"I figure her to be of Italian descent. Her tits aren't real big, but I've fallen in love with her tight little ass."

Nodding, Ed offered, "I'll bet she's been ridden a mile or two."

"Ya think? She looks like she's been ridden hard and put away wet."

Lee frowned with disgust.

"I'm not interested. You can take my turn."

Thirty minutes later Gina bade them farewell after three quick rounds. As she slipped past Lee in the hall, she hesitated and then stopped.

"I'll get you the next time I drop in. You and your friends come here often?"

"We'll probably be here most every weekend."

"Next weekend, then. Same time, same station?"

Eagerly, Bill confirmed, "Count on it."

The *USS Tanner* arrived in port on January 8, setting sail for Norfolk a month later. On the mess deck after sea detail had been secured, Auggie nodded head reflectively.

"Working on the deck force is getting a little old. How'd you manage to get assigned to the Ship's Office?"

"I applied for it. Why don't you and Bill apply? There's a couple of openings."

"Sure beats chipping paint. We're kinda liking not having to stand watches, aren't we, Lee?"

"You bet. Ed and I've kinda missed you not being able to hit the beach every night. Ah, how's everything going with Gina?"

Bill shook his head, somberly reporting, "She's lived a hard life. Do you realize her old man forces her to stay out on the streets until she reaches her quota?"

"How many tricks does she have to turn?"

"She told me she does twenty tricks a night."

"Ya, and if her old man's not beating her because she doesn't meet the quota, the cops are busting her."

"It's what they're supposed to do, isn't it. She is street walker."

"They don't bust her for tricking. They nail her when it's her turn to do some time in the slammer. Seems like all of New York's finest is on the take."

"Why you continue to see her? How much fun can it be to always have to pay for it?"

Bill frowned, methodically hiking up his sagging dungarees.

"What are we supposed to do?"

"Look for someone who's not prowling the streets."

Bill protested, "Gina's trapped. If she doesn't work, she gets beat by her old man. If she works, she ends up in the slammer for soliciting. I doubt she's ever done anything except turn tricks."

"Yeah, and someday, she'll probably end on some back page. Just another statistic."

Lee paused, disgustedly shaking his head. "You and Auggie aren't helping her situation one iota, Bill."

OVER THE BOUNDING MAIN

Uncle Sam deployed a large gray flotilla resolved to protect and defend, mustering seafaring men to cautiously guard the peace. Join the Navy to See the World, Sailing the Seven Seas! Ah, the life of a sailor, adventure, travel, and a girl in every port.

Better get a letter off to Sandy. Yeah, might be nice if I congratulated her for being promoted. Pretty impressive, now she's the chief prosecutor for Marion County. Who knows where we're going after we leave Norfolk or how long it'll be before I can write to her again? The rumor is we're gonna do some surveillance work in the Barents Sea. Strange, I actually thought the Tanner *was a coastal geodetic survey ship. Was I surprised to learn her mission wasn't to study plant and marine life while mapping the ocean's floor, or what?*

The gravity of the *Tanner's* mission became much clearer when the division petty officer announced a change in policy for access to the captain's office.

"From now on, the captain's office is off limits unless you have a top secret clearance. Along those lines, when we get to Norfolk, we're gonna muster aboard some additional crew. Normally, your office would handle the paper work, but the records of these people are being kept in the captain's office. I doubt you'll be affected much because it's highly doubtful that you'll even know they're on board."

Lee frowned.

"Sounds kinda weird to me. Any reason why we're not going to handling their records?"

"Let's just say that any information about these people is strictly on a need to know basis."

Tom White, a shipmate who worked with Lee in the Ship's Office leaned back in his chair and chuckled as he took a deep drag on his cigarette. "Hush, hush, huh?"

Lieutenant Morgan nodded. "You might say so."

The hint of a mystery delighted Tom, provoking him to chuckle with delight.

"If we can't ask about the new personnel, is it at least okay to ask why we're picking up all the electronic surveillance gear in Norfolk?"

"Who told you about the electronic gear?"

Tom's eyes danced with delight.

"I've got my sources."

"I'd forget you heard anything about our ship taking on surveillance gear if it were me. The less you know, the better off you're gonna be."

"I'd put money on the *Tanner* heading back to the Barents Sea as soon as we leave Norfolk."

Tom's Cheshire Cat like smile defied denial.

"Just wait and see. The last time we went up there, we went to Norfolk first."

"Think what you want, Tom, but, personally, I think you have more important things to worry about. I'm sure we'll find out everything we need to know as soon as the Captain is ready to tell us." Grinning, Lieutenant Morgan added, "Besides, we could do worse than go to Key West."

Tom frowned. "What's down there?"

Lee chimed in, "Cuba. The last time I checked it was only ninety miles south of Key West. If what you say about the surveillance gear is true, I wouldn't be surprised if we're not headed down there to listen in on all the Cuban transmissions to see what old Fidel has up his sleeve."

Tom frowned, the wheels of his over active imagination working overtime.

"Hmm, I never thought about Cuba, but, now you mention it, it makes a lot of sense. The Navy's been watching every move they make for quite some time now. Just because the Russians say they're removing the missiles doesn't mean squat."

"This is exactly how rumors get started. All I did was mention Cuba and all of sudden we're going down there on a mission. I think we need to worry about our jobs in the Ship's Office, Tom, and leave the operation of the ship to the captain."

When the *Tanner* docked in Norfolk, Lee left the ship to meet Bob.

Bob sure has a cushy assignment. How tough would it be having to play golf with Captains and Admirals? Hmm, he wrote he was going to introduce me to some guy by the name of Georgan Waitman. I guess he's from New York and supposedly knows lots of women up there. Maybe he'll give me a name or two. Anything would beat the hotel scene Bill and Auggie seem so attracted to.

After introducing Georgan to him, Bob as he normally did, assumed command of the gathering.

"Boys, we can either go out on the town and get laid, or we can find some quiet little bar and share a few stories. What's it gonna be?"

Lee chuckled, shaking his head in amazement.

"I guess Bob forgot to tell you, Georgan, his track record with women isn't very good. Maybe we'd better hit the bars and forget all of this fantasizing about getting laid."

"Yeah, right. Lee hasn't had enough time to train me in the fine art of chasing women. I guess drinking a little beer would be best, though. I'd hate to embarrass a Bearcat by showing him how it's done."

"Oh, sure, Bob. Waitman, can you believe this sorry Pioneer actually passed up a freebie on the train coming back to report for duty? As I remember, Lisa from Seattle offered to give it up. What was it you were short of, nerve or cash?"

Waitman grinned, enjoying the debate.

"Maybe he's not trainable, Lee. Anyway, before this gets out of hand, I want you two know that I didn't plan on watching a tennis match. Um, I suggest we do the bar scene. If nothing else, I'm sure we can keep each other entertained sharing stories about our college days. Speaking of which, I still have a few contacts in New York. One in particular stands out in my mind. Would you like her name and phone number, Lee?"

"I've got girl back home, but if you want to tell me about her, I'll listen."

"Variety is the spice of life, Lee. Course, ah, if it's serious with the girl back home, ah, I don't have to share."

"I don't know what harm it'll do to find out a little about her."

"Well, she lives in upper Manhattan. I got to know her very well while I was going to NYU."

"Just good friends?"

"Very good friends."

Bob chuckled, eager to again ignite the debate.

"If she goes down, she'll have to show Lee how."

"Do I have to remind you again of Lisa from Seattle, Bob?"

"Hostile! You guys have been at each other since we hit the beach. Um, as I was saying, Lee, she's something else again."

"I guess taking her out for a drink wouldn't hurt."

"Just so you know, she was engaged." Grinning knowingly, he quickly added, "But it didn't stop her. We still went out."

"And?"

"It wasn't exactly casual. Fact is, it was starting heat up pretty good until I found out her fiancée was involved with the Mafia."

"You're kidding. Why would I want to get involved with the criminal element? I'm not so hot to trot that I'd go looking for someone to put out a contract on me."

"I wouldn't have even mentioned Margaret to you except I heard she and the Mafia guy parted company."

"Sounds like she has a history of playing around."

"Ah, you might say that." Nodding his head, he continued, "She's been known to play the field. However, except for yours truly and the Mafia guy, most guys don't last very long with her. Ah, I hear the reason she and the Mafia guy called it off is because he's doing the daughter of his boss. I think it's a command performance, if you get my drift."

"Was it love with the Mafia guy, or did she just like to fuck him?"

"I think she likes to fuck, period. Anyway, I'll write her name and phone number on this coaster. You can do what you want with the information."

Georgan smiled slyly.

"If you treat her right, she'll probably let you to call her Peggy, and believe me, she is one good-looking woman with a body to match. Um, you might mention I gave you her name. Ah, it just might open the door for you."

Three weeks after pulling out of Norfolk, the *Tanner* had completed its mission and was headed back to New York for a visit to the dry dock.

I sure am glad when we get this luxury liner docked and I can hit the beach. Constantly being on the lookout for Cuban aircraft and PT boats got a little old. Talk about tedious! About the only interesting part of patrolling outside Havana's harbor was listening to the Cuban music and the constant haranguing of some of the political radicals on the radio. Be nice to be able to tell Sandy and my folks what I doing and where I am again.

Tom White leaned back in his chair after finishing some separation orders for one of the crew scheduled to leave the *Tanner* when they docked.

"See, I told you so. The Navy department wants to see if this old Liberty ship is up for a return visit to the Barents Sea. I figure we'll be heading back up there in six months, tops. Um, any of you making the trip from Hoboken into downtown Manhattan after we dock?"

Lee nodded, quickly cleaning up the mess on his desk.

"I'm going to visit a fraternity brother and his wife. Bill and Auggie are probably headed into Manhattan."

Lee turned towards Auggie.

"You and Bill going to look up Gina when you hit the beach, Auggie?"

"Nah, Bill and I are probably gonna save a visit with her for the weekend. We're just gonna find some bar and enjoy a beer or two. Um, the last bus back to the ship from Brooklyn is at 2300, right?"

Tom nodded, smugly lecturing, "I wouldn't plan on missing it, either. Cabs cost a fortune. Being late to muster will be costly, too."

Bill methodically turned away from his typewriter.

"What's Ed gonna do?"

"He's gonna stay aboard ship."

Auggie nodded, staring out the porthole.

"Maybe waiting until we move this bucket of bolts back to Brooklyn this weekend isn't such a bad idea."

After a most enjoyable dinner, Lee left the home of Peter and Mary Welk at eight-thirty. Outside the bus station in Brooklyn, he stopped at a telephone booth.

Hmm, I think I'll take Georgan's advice and give Margaret a call.

He took a road worn coaster out of his wallet.

Well, the worst thing that Margaret Gilland can do is to tell me she's not interested in going out for a drink. I doubt Sandy's gonna be sitting around gathering moss. Waiting for me for eighteen months is a long time. Course with her investigating the Mafia family, she could be kept pretty busy. What was it she said, the head of the crime family just recently relocated Seattle? Strange, Mr. Knight and his wife just moved to Seattle.

He reached for the telephone.

It was really a nice dinner Pete and Mary served. I'll have to look them up again sometime. Maybe I'll get them some tickets to a play or something. Sure is nice of them to give me an open invitation to come to dinner. He released his finger from the dial after dialing the final number and waited for the dial to complete its rotation. After a momentary pause, the phone started to ring.

"Hello?"

"Is Margaret, Margaret Gilland there?"

"May I ask who's calling?"

"You may. Ah, my name's Lee Grady. The name doesn't mean anything to you, but if you'll give me about three minutes, I'll try to explain how I came by your number. Ah, hello? Hello, are you still there?"

"You said three minutes. The clock's running."

"Right. A few weeks ago, I met a friend of yours in Norfolk, Virginia, ah, Georgan Waitman."

"My goodness. Of course, I know Georgan. In fact, I know him very well. How is he? I haven't heard from him in ages."

"He's fine. He indicated you and he were pretty good friends. Um, he suggested I give you a call."

"I'm glad you did. What's he doing now? The last time I heard, he was thinking of joining the Navy."

"You heard right. Um, he's stationed in Norfolk. Yeah, well, he gave me your number and suggested I give you a call."

"Do you always repeat yourself?"

"I did, didn't I? Ah, well anyway, he's a real party animal. I met him while my ship was in Norfolk. A friend of mine introduced us."

"Same old Georgan." Giggling, she offered, "Sounds like he hasn't changed much. He and I used to tie one on occasionally. Maybe he told you we went to school together at NYU." After a short pause, probed, "So, what's on your mind? Your three minutes are just about up."

"I was hoping we could get together. Ah, before you say no, please hear me out."

"I suppose this is going to cost me another three minutes. Oh, well, go ahead. What's another minute or two?"

"Okay, then, ah, I thought maybe I could come by your apartment and meet you."

"Oh, you did, did you? And then what?"

"Well, when I knock at your door, you'll answer it, right?"

"I would think so."

"Just to be on the safe side, you could make sure it's chained so you don't have any worries about me breaking in. If you'll allow a few minutes for us to talk and look each other over, ah, well, if you don't want to invite me in, you can just tell me to hit the road."

"Is this another three minute deal?"

"I was hoping for a little more then three minutes. Oregonians usually need longer for something as important as an introduction."

"Westerners are supposed to be quick on the draw, aren't they?"

"No, it's the eastern folks who always seem to be in a rush."

"You don't say. Hmm, so suppose I think you're okay after our three-minute visit, what do you propose, then?"

"You could always invite me in, or so we could go some place for a drink. Of course, we could always just stand there and look at each other."

"Hmm, I don't know. When do you propose we meet? My schedule is booked on the weekends."

"Fridays too?"

"Uh-huh."

"How about tomorrow?"

"Hmm, Tuesday. I suppose, ah, but you don't even know where I live."

"If you tell me and I can't find it, you're off the hook."

"Hmm, okay, I'll give you the three minutes you requested, but no promises."

"What time?"

"How does seven sound?"

The next day after leaving the bus from Hoboken, Lee boarded the subway headed for upper Manhattan after glancing at the address he had written on the coaster.

Hmm, Margaret lives right in the heart of the Germantown district, same place I went with Laura when I visited New York with Bob and Jim. Hmm, I wonder how Laura's doing? I was supposed to look her up, but I can't even remember her last name.

Trying to arrange a three-minute meeting with Margaret was like trying to catch a fuckin' eel. After what she put me through, tonight had better be good. Georgan better not have put me on. She did sound pretty good on the phone, though. No matter how good she is, she's still just gonna be a person to spend time with. God, I miss Sandy.

He again studied the address of Margaret's apartment that he had written on the worn coaster.

Hmm, she doesn't live too far from Peter Welk. If I leave at eight-thirty or a quarter to nine, I should make it back to the bus station in plenty of time.

A slow, sauntering ten-minute walk from the subway stop gave him ample time to ease his nerves, ample time to plot out his three-minute spiel.

Hmm, the plate below the doorknocker says, M. Gilland.

He checked his watch.

Good, I'm a little early. Plenty of time to adjust the tie and take off this stupid hat. I'll be damned if I want to look like a typical Navy lifer. Well, here goes nothing.

A moment after he knocked, the doorknob turned, and the door slowly opened. Partially exposed in narrow opening, a figure studied him with caution behind the restricted access allowed by a security chain.

"You must be Lee."

"And you must be Margaret. I hope the clock hasn't started running already. I was gonna petition for an extension, anyway."

She opened the door a little wider.

"I think a little time to warm up is only fair."

"Thank you."

Interviewing her presentation, he noted her light effervescent green print dress with spaghetti straps concealed the shapely figure of someone with unassuming confidence. Cosmetic assistance, if any, offered only a mild contrast to her freckled, chestnut complexion.

Hmm, long auburn hair and even white teeth. Damn, the perfume she's wearing is starting to give me ideas. Georgan was right. She's flat out beautiful. What a body!

"If all you're going to do is stand there and look at me, I might just as well let the clock start running. Hmm, I see you wore your uniform. Ah, just so you know, I'm not real fond of the military."

"The military in general or specific members of the military?"

"There's a difference?"

"I almost didn't wear mine, but, well, a three-minute meeting made changing seem like such a waste."

"Yeah, I suppose for a three-minute introduction it would hardly be worth the effort."

Grinning, he inquired, "How am I doing so far?"

"Do you need an evaluation?"

"I suppose not. Ah, I was just wondering if you're disappointed."

"The reputation of military people repulses me."

"What are you referring to?"

"Everything I've heard or read portrays them fucking anything daring enough to move." Frowning, she added, "Then, to make sure they did it right, they nail them again just to insure they didn't overlook 'em. Is that how you operate or are you as different as you appear?"

"I'm more discriminate. I only nail every other one."

Grinning, she nodded.

"Good for you. Ah, no, I suppose I'm not disappointed. Then again, I guess I didn't really know what to expect. I, ah, I've just never gone out with a sailor before. I'm not so sure I want to start, either."

"I think I still have a minute. Think it's enough time to convince you I'm not the typical sailor?"

"You could give it a try."

"Okay, then, ah, I come bearing truth, and an apology for not wearing the attire of the common man on the street. However, I thought you should see me as I am before I step into the phone booth to assume another identity."

"With or without a cape?"

"Don't forget the 'S'. Um, I can't help it if I'm in the Navy, but if it's too big a deal, I'll leave without argument."

"You must not have criminal intent, otherwise you would have torn down the door by now." Grinning, she suggested, "I haven't gotten my money's worth, so, it looks like you've earned an extension." Looking at her watch, she petitioned, "How's another minute sound? I'm trying to overcome a bad habit of mine. Judging a book by its cover isn't always the best policy."

"Georgan didn't lie. You're all he said you'd be, and more. I, ah, I didn't expect to be so lucky."

"Don't get your hopes up. Uh-huh, maybe you should tell me exactly how Georgan described me."

"Well, he told me I would be impressed. He mentioned you were intelligent and good-looking."

"Yeah, yeah, I know. All the stuff he knew you wanted to hear."

"Maybe so, but he didn't do you justice."

"Oh, sure. I'll bet you say that to every girl you meet for the first time. Where did you say your ship's located?"

"I didn't, but we're located in Hoboken."

She snorted, wrinkling up her nose, and shaking her head.

"I can't believe you said Hoboken! It's the armpit of the world, Mother Nature's sewer."

"We're only going to be moored there until Saturday, then we're moving back to the Brooklyn Naval Yard."

"I'll bet you'll be glad. How do you get from here to there and visa versa?"

"By bus, subway and good planning. Um, the last bus leaves for Hoboken at eleven."

"Oh, so, you didn't plan on staying the night?"

"Not hardly. I can only stay until about nine-thirty."

"Would you like to come in?"

She smiled, shyly ducking her head.

"I think I've tortured you long enough."

"I passed the test?"

"I'm still evaluating. I agreed to meet you out of respect to my friend, Georgan. So far, nothing I've seen bothers me."

Fidgeting with the chain on the door, she finally worked it free.

"Extending your visit shouldn't hurt anything, should it?"

"Probably not. I'm not really dangerous. Persistent, but never dangerous."

"Maybe your powers of persuasion are better than you thought. Then again, sometimes good things take a little longer."

"I doubt a couple of hours is enough time to stir up too much trouble."

She motioned towards the couch.

"Make yourself comfortable while I get us something to drink. Ah, what would you like? I have beer, wine, or coffee. It's your call."

"What about you? What are you having?"

"I think I'll have some wine."

"I'll have a beer, then."

"Why don't you tell me a little about yourself. So far I know you're in the Navy and a friend of Georgan's. Surely your resume doesn't stop there."

"I'm twenty-three, graduated from Willamette University in June, and, ah, right now, I'm just trying to fulfill an obligation."

"Where in the world is Willamette?"

"Salem, Oregon. Um, Georgan told me he thought you broke off your engagement."

"Kinda. A lot's happened since I saw Georgan last."

"Well, I'm all ears."

"Um, my life's kinda complicated. The engagement's on the shelf for now, but we're still seeing each other. Still wanta hear more?"

"Sure."

"My resume would point out that I'm twenty-five. Two years ago I graduated from NYU, majoring in dental technology. Um, I was on the extended study plan."

"No big deal, it took me five years, too."

"About the engagement, ah, did Georgan tell you anything about the guy I was engaged to?"

"He said your 'Ex' was in the Mafia."

"And you still called? When Georgan found out, he couldn't run fast enough."

"Are you still seeing him, um, the Mafia guy?"

"Yes. Um, are you interested in the sorted details?"

"If you want to share."

"Yeah, maybe if I have to explain it to you, I'll understand what's going on myself. Um, Vinnie, my 'Ex', and I were seriously thinking about getting married when his boss, the head of the family, called in a favor."

"A favor?"

"Yeah, it seems the Don's daughter got herself in a motherly sort of way. The Don told Vinnie to marry her so the name of the family wouldn't be tarnished."

"And Vinnie obliged?"

"With an understanding. Ah, he and the Don's daughter will be married, but Vinnie has certain freedoms."

"Uh-huh. He still gets to play house with you, right?"

"He drives up to my parent's home in Connecticut every weekend to see me."

"Do your parents know?"

"They think we're still engaged. They don't know anything about Vinnie being married."

"Where does Vinnie live?"

"He runs the family business in Boston. The Don and his wife moved to Seattle to run the operation there." She smirked, inquiring, "Is my life complicated enough for you?"

"How does Vinnie pull off seeing you every weekend?"

"He has an understanding with his wife. Ah, they're married in name only. He thinks when the baby comes he'll be able to get out of the marriage."

"And then you plan to marry him?"

"I don't know. I'm not liking the mess he's in, even though he didn't have a choice."

"But you still fuck him on the weekends, huh?"

"Persistent, aren't you? Yeah, I'm still seeing him and doing him on the weekends." Frowning, she pleaded, "Why don't we just drop it. I'd much rather discuss your life than mine."

Lee shook his head, starting to rise from the couch.

"Tonight might not have been such a good idea. Um, I probably should go. Just so you know, ah, I'm kinda involved with someone back home."

"So, we both have someone in the closet. Why don't you sit down while I turn out the kitchen lights? They're a little annoying, don't you think?"

"I suppose, but, ah, I thought I told you this wasn't a very good idea."

"I'm a poor listener. Why don't you douse the pole lamp in the corner?"

She checked the clock in the kitchen.

"Hmm, it's eight. We still have an hour or so, right?"

"I suppose."

"Well?"

I don't know about this. I'd love to hold her and explore the fire that seems to be glowing beneath her beautiful exterior. Damn, her body is phenomenal. She reminds me a lot of Judith, the dynamo I met at the inn in Maine. Margaret seems to want to move beyond the introductions and get right down to business. Tempting. I'm not gonna do anything to mess up what I have going with Sandy, though.

She turned off the light and started to feel her way back to the couch. When she returned to her station on the couch, their relationship no longer was at arm's length.

"So what else did Georgan tell you?"

"Nothing. How about telling me a little more about Vinnie."

"Vinnie and I enjoy fucking each other. I suppose we might even love each other."

"Vinnie, what's his last name?"

"Trabago. You need to know all the details, don't ya?"

"If you're still love with Vinnie, why me?"

"Let's say that I'm attracted to you."

"I'm not exactly excited about messing around with some Mafia guy's lady. Besides, it's a little sudden isn't it?"

"Time's on my side. You have to leave in a few minutes, don't you?"

"Yeah. You did say it was a little after eight, didn't you?"

"When I turned off the lights, it was eight-fifteen. We still have an hour. Unless it's just my imagination, you want the same thing I do. Um, you'd like to kiss me, wouldn't you?"

"The thought has crossed my mind."

"Afraid the sailor image will work against you if you try?"

His face started to redden.

"Well, you brought it up."

She leaned forward and kissed him tenderly on the lips. When their lips parted, she looked at him momentarily. Suddenly, she again eagerly sought out his lips, passionately offering herself to him.

When their lips parted, she whispered. "I like the way you kiss."

"I like the way you kiss, too, Margaret."

"Why don't you call me Peggy. People I really care about call me Peggy."

"Peggy, I'm not going to lead you on with a bunch of bullshit. The gal back home, I love her and have every intention of marrying her."

"But?"

"The problem is, I'm also attracted to you."

Like a multicolored chrysanthemum bursting in mid-air during a Forth of July celebration, warm colorful rays of light radiated throughout the room. Sparks flew! A sudden warm glow invaded his gut and spread quickly throughout his body. Her intoxicating presence did not allow him to defend against a growing desire to invade her space.

"I want you."

"I want you, too. Touch me?"

He bent forward, greeting her responsive lips. After a long embrace their mouths parted. He looked at her and then kissed her again. She responded passionately, pressing her breasts against his chest. After the exchange their lips again parted.

She gasped. "God, I love the way you touch me."

Eagerly she again greeted his lips, offering no resistance to his advances. Their lips parted.

Sighing, she whispered, "Wouldn't it be better if this darn dress weren't in the way?"

"You read my mind, but we have a small problem."

She exhaled loudly.

"Right. You have to leave. Damn, it was just starting to get interesting."

"Just remember where we are. We can take up where we left off, next time."

"Tomorrow?"

"No, getting here and back is too much of a hassle. Let's consider next week? We should be relocated in Brooklyn by then. Um, would Monday work?"

"How about Monday, Tuesday, Wednesday, and Thursday?"

She smiled, toying with his tie while waiting for an answer.

"Ah, is there any chance of Vinnie finding out?"

"No, why should he? Do you plan to tell your girlfriend?"

"Touché!"

"When you come to see me Monday, why don't you plan on coming for dinner?"

"What time?"

"How soon can you make it?"

"How does five-thirty sound?"

She nodded. Suddenly she rose from the couch and headed for the kitchen.

"Time check."

She turned on the lights.

"Oops, it's nine-forty."

"Damn, I'd better get going. Five-thirty on Monday, right?"

"Better still, why don't you pick me up where I work? I get off at five. We can wander around and see what mischief we can get into before we come back here for dinner."

"Where do you work?"

"My office is at 5016 Fifth Avenue. Just wait out front for me on the street and enjoy the sights. By the way, lose the uniform and wear your common man on the street outfit. I'm not going to hold it against you, but I'm not real fond of the Navy."

The subway ride back to Brooklyn was a time to celebrate and a time to reshape his thoughts about the next few months before the *Tanner*'s next deployment.

Sure glad it's the last time I have to make this trip. Monday when I come to see Peggy, I might even consider staying all night. Georgan was right, she's quite a woman. Seems like she's all he said and maybe a bit more. Ah, I still need to a little more clarification about Vinnie. The Mafia connection's a little scary.

He closed his eyes.

I'm starting to have problems even remembering what Trish looked like. Strange, Diedra didn't sound like she knew about the accident. I wonder what she meant when she said Trish's dad would never have let us get married?

A smile appeared on his face as his mind recalled a letter Sandy had written.

She wants me to call her this weekend. I wonder what the important news is that she has to share with me? Think I'll call her Saturday after we dock in Brooklyn.

When the *Tanner* was once again moored its berth across the East River from the New York skyline, Lee headed for a telephone booth outside the gate to call Sandy.

"This is the operator. Will you accept a collect call from Lee Grady?"

"Yes, yes I will."

"Go ahead, please."

"Hi there, Counselor. How's life in the justice system?"

"Oh, Lee, I've missed you so much. Thank you for calling. It's so good to hear your voice."

"I've missed you, and it's good to hear your voice, too."

"It's been so long since I've heard anything from you. Are you okay?"

"I'm fine. After I wrote to you in Norfolk, I couldn't tell you anything because we were on a mission, ah, top secret. Did you get my letter telling you we were docking at Hoboken?"

"Yes. When I got it, I decided to have you call so I could hear your voice. God, I've missed you."

"I've missed you, too. You had me worried. Your letter sounded kinda mysterious."

"Honey, the letters have been good, but there's nothing like being able to hear your voice. When do you get to come home on leave?"

"We're pulling out at the end of May or early June for another mission. I'm hearing we'll be gone until December."

"December, do you mean other than a few letters we won't be able to talk with each other, until then?"

"I suppose you're right. Once the mission begins, the only way we'll be able to talk is through the good old postal service. Ah, when the mission's over, I'm taking a thirty-day leave."

"Music to my ears, ah, it wouldn't do any good to ask where you're going, would it?"

"Honestly, I'd tell you if I knew. All I can tell you now is what I've heard through the rumor mill."

"Well, I guess a rumor is better than nothing."

"Have you heard of the Barents Sea?"

"North of Russia?"

"Yep. Ah, once we pull out, the mail isn't going to be real regular. I'll write all I can, but you should be prepared for the worst. We may not hear much from each other until I get home for leave."

"How depressing. Ah, I've got news, too. It looks like the work I'm doing for the U.S. attorney's office is going to be a little larger job than I anticipated. Ah, I'll be heading up the case, the one with the Mafia connection I told you about. So far, there's really nothing to report. It looks like this could be a long drawn ordeal. Anatolia is sly like a fox, but I'm working on a few angles. We'll get him. It's just a matter of time."

"Will the work affect your position in the DA's office?"

"Not for now. It's a real opportunity."

"Good. It's what you've always wanted."

"It is, but I've always wanted you, too. God, I've missed you. I love you, you know?"

"I love you, too. Not being able to see you is really hard. I'll sure be glad when our on again, off again relationship comes to an end and we can be together permanently."

"Me, too. Ah, what do you think about calling me every weekend until you have to go on the cruise? Weekdays are so hectic with my schedule and all, you'd probably never get hold of me."

"Are you kidding? I can't think of anything better. Hearing your voice makes all the difference in the world."

"It's settled, then. Plan on calling me either on Saturday or Sunday. Ah, one other thing. I know it would be unfair to expect you to sit around for the next eighteen months twiddling your thumbs, but I was wondering."

"About what?"

"I don't want anything to mess up our future together."

"You don't have anything to worry about, Sandy. I don't plan to just sit around on the ship until my time is up, but I'm not going to let anything or anyone stand in the way of us being together when this service gig is all over."

"Just what I wanted to hear. I love you so much. I don't want to take any chances of losing you, and, ah, from my end, you should know there's nothing to worry about, okay?"

"Okay."

"I haven't heard any specifics since you were in Norfolk. Mind filling me in?"

"Not much to report. We sailed to Key West and then went on station off the coast of Cuba. I've been told we were monitoring Cuban communications to ensure nothing strange was going on. You don't know how close we came to getting involved in something I don't want to even think about."

"I know. The news reports were pretty gruesome. Thank God President Kennedy was able to call the Russian's bluff."

"Other than a few weeks of trolling around in the beautiful southern Atlantic, nothing much happened. Oh, yeah, we ran into a hurricane on our way back up the coast to Hoboken."

"Only a hurricane? I would call running into a hurricane a real big deal. What was it like?"

"When it hit, I was asleep in my bunk. I guess the backhand of a swell hit the ship broadside. The next thing I knew, I was face down on the deck. You should have seen our office. We forgot to secure it when we secured for the night."

"I'll bet it was a mess."

"Yeah, typewriters and papers were strewed everywhere."

"What was it like sailing into New York? I hear it's spectacular."

"It is. You know what it's like when you've been away from home for awhile, and you catch the first glimpse of your home?"

"I sure do."

"Well, triple the sensation and you might be close. Seeing the Statue of Liberty and the New York Skyline was almost spiritual. It was the most awesome panoramic sight I've ever seen. From my station on the ship, every visual affect was like viewing something through a spherical orb. One moment I was looking at the New York skyline with the sea water from the channel rushing out to greet me, and the next I was looking at Ellis Island and the green hued lady standing there welcoming us. When I saw her right hand holding her torch, and her left hand cradling the tablet, I think I know how the immigrants might have felt. For them, it might have seemed like God was proclaiming a new beginning."

"Sounds to me like you've fallen in love with New York?"

"I have, but as much as I love it back here, my heart's still back home with you. Sharing a life with you is what I really want."

"Ditto. Ah, how do you get around? Do you use the bus, take the subway, take a cab, or what?"

"I had to use the bus while we were docked in Hoboken, but, other than walking, I use the subway. Do you realize how narrow Manhattan Island is?"

"On the map, it looks like everything can be reached by walking."

"It is if you restrict your walking path to an east-west direction."

"How much is the subway fare?"

"Can you believe, fifteen cents? Ah, Sandy?"

"Yes?"

"I want you to know, I plan to knock on your door when I get out of this outfit. I'm hoping you'll answer."

"Count on it. Everything I do, and all my thoughts are focused on the moment when you come calling. I love you, Lee. Don't forget to call me next weekend."

THE SPY GAME

The cold war was a game of cat and mouse. A stare-down in Cuba, Russian arms parades, American spy planes, atomic weapons testing, and covert missions tested the resolve and patience of the major players as the world teetered precariously close to the cusp of armed conflict, nuclear war. Neither side knew who was the cat or the mouse.

The dress code is sure different in New York. Everyone seems to be wearing three-piece worsted suits, a topcoat and hat. It sure is a lot more casual back home. Man, there sure are a lot of people parading up and down the street. Wonder where they all are going? I suppose they must work the high rises, probably in advertising or something of a similar nature. Hmm, I wonder what those people passing by would be able to share if someone took the time to interview them? They're probably thinking I'm underdressed. Then again, they probably don't give a shit. They're all in too big a hurry to even notice I'm standing here.
"Pardon me, but would you happen to have fare for the subway?"
A familiar voice interrupted his preoccupation with the scurrying throng of people in a maddening hurry to go somewhere, nowhere.
Lee reached into his pocket, extracting the contents.
"Hmm, all I have is a token, will that work?"
"You look nice. I wasn't sure what to expect, but I'm glad you lost the uniform."

"Thanks. I've been counting the minutes for this moment to arrive. I thought about you all weekend."

"You've been in my thoughts, too."

"How was Connecticut?"

"Fine."

She grabbed his hand and started leading him up the street.

"I know I told you we might look around for some mischief to get into before we go home and have dinner, but I've reconsidered."

"Oh, you have, have you? Um, what's for dinner?"

"How does steak, potatoes, and salad sound?"

"I'm impressed, but you needn't to go to so much trouble."

"What trouble? I have to eat anyway. Besides, I enjoy cooking, and all bragging aside, I'm pretty good at it."

"You know the way to my heart. Ah, since we're not gonna explore, could you possibly be looking for a different kind of mischief to get into?"

She chuckled, poking him in the ribs.

"You'll never find out if we just stand here and talk about it."

At the Germantown stop, the train jerked to a halt. Quickly vacating their seats, they melted into the crowded rush of commuters impatiently pushing and shoving their way towards the exit. Finally on the street above, they merged with a diminished mob on the narrow pedestrian lined sidewalk.

"Did your ship get relocated, ah, at the Brooklyn Naval Yard?"

"It did. It sure is an easier commuting to your place now that we've moved."

"How long does it take you to get here from your ship, now?"

"Counting the stop at the YMCA to change, less than an hour."

"What if you didn't have to stop at the YMCA to change?"

"Hmm, I guess it'd take about thirty to thirty-five minutes."

"Hmm, if you were to keep some of your clothing at my place, it would be much more convenient, wouldn't it."

"Not a good idea."

"Why?"

"What if Vinnie were to find out?"

"Are you kidding? He never visits me in New York."

"Why?"

"Let's just say, he watches his back pretty carefully. Besides he has a home in Connecticut. His parents and mine live there, so it makes everything pretty convenient."

"Does his wife know?"

She shrugged, disinterest etched on her face.

"I doubt she cares. It's not like they're sleeping together, ah, a marriage of convenience. Ah, at last. When I'm in a hurry, the walk from the subway station seems to take hours."

She put her key in the lock and turned it. Turning towards him, she smiled as she pushed the door open.

"You know the way, why don't find a seat and enjoy some television while I start putting dinner together? It shouldn't take long."

A moment later she reappeared in the doorway to the kitchen.

"I'm sorry, I wasn't thinking. If you would like to look around, feel free. Um, could I offer you anything to drink?"

"I'm fine, thanks. Ah, are you suggesting that it's okay if I explore your digs?"

"I don't have anything to hide."

The sound of the refrigerator opening and her mumbling provoked him to check out the activity in the kitchen. Sensing his presence, she turned towards the doorway.

"What?"

"Nothing, I was just watching you. You were mumbling."

"I always talk to the carrots." Waving her arms in frustration, she protested, "Excuse me for not being a very good hostess, but please don't stand there watching me. It makes me nervous."

"Relax, Peggy. I was just peeked in while I was casing out the joint. If there's anything I can do, just holler."

He returned to the living room and started looking at some of the decorations, fingering a couple of novels on her bookshelf. Suddenly his attention shifted to an antique wooden box beside the hi-fi cabinet. Close inspection revealed it contained her record collection.

"Hmm, nice. You've got a little bit of everything in your record collection. Ah, if it's okay with you, I need to use the facility."

"It's okay to say bathroom. Ah, you'll have to excuse the underwear hanging on the shower bar. I can't see wasting the quarters to use the clothes dryer."

"Aren't you afraid of revealing secrets."

"Not unless you get your jollies off by looking at a girl's undies."

"Not really. I prefer to see them when they're being modeled."

Hmm, let's see what the bathroom is like. A bathroom can sometimes tell a great deal about a person. Hmm, gotta hand it to her, it sure is neat and orderly. Wow, She must be a collector. What a neat set of miniature painted porcelain dolls!

He quickly spun about in the center of the bathroom to give it one more tour of inspection.

Good, no sign of an early morning rush. She seems to be a good housekeeper.

After washing his hands, he opened the bathroom door and stepped into her adjacent bedroom. Her bed was located in the middle of an outside wall. The window above the bed was neatly framed with decorative curtains. Her bed was neatly made with a multicolored cover and two large pillows showcasing a beautiful China doll. To the right of the bed, closed mirrored sliding doors concealed the contents of her closet. A clock radio alarm clock, telephone, and family portrait rested on the nightstand near her bed.

Hmm, queen-sized bed. If I'm lucky, we just might make use of it a little later.

"Is this a picture of your parents on the nightstand?"

"Yeah, it was taken when I graduated from high school."

A moment later he took a seat on the couch in the living room. Reintroducing himself to the hi-fi, a television set, and a large overstuffed chair, he noted several pictures and a large painting of the New York skyline neatly positioned on the wall. Two corner pole lamps, one table lamp located on an end table, and a long coffee table with some magazines neatly piled on it, helped finish off the room decorations. He got up, and walked over to peer through the large opening cut in the wall separating the kitchen from the living room.

"Just checking to see how you're coming with dinner. Anything I can do to help?"

She smiled as she busily continued to slice the tomato wedges for the salad.

"You could pour the wine, if you'd like." Motioning towards the left end of the counter she stood behind, she revealed, "The wine and opener is on the counter. I'm thinking white wine tonight, how about you?"

"Fine with me. What glasses do you want to use?"

She nodded towards the cupboard.

"The glasses are on the left side. When you're done pouring the wine, would you please put some music on? I'm going to take a shower while the potatoes are baking."

"Do you want me to do anything else while you're showering?"

"Thanks for asking, but everything is under control."

She took off her apron, folded it neatly and put it on the counter.

"Just sit back, relax, and enjoy your wine. I'll be back in a jiffy."

He removed the cork and poured the wine, setting her wine on the dining table.

"What kind of music do you like?"

"I'm in the mood for some Mancini or Coniff."

During dinner, the dim candlelight cast darting shadows as they scrutinized each other from across the table. Her appealing presence, unspoken invitations transmitted through suggestive glances, quashed his appetite. Thoughts of Sandy stirred a feeling of uneasiness within, inducing him to toy with the unfinished meal.

Frowning, she inquired, "Is there something wrong with my cooking? You've hardly touched your food."

"No, it was fantastic."

"Well, then, what's the matter?"

"Nothing's wrong, I'm just full."

"I'm not trying to pry, but my relationship with Vinnie is bothering you, isn't it?"

"A little. I was also thinking about Sandy. Ah, I don't want to get involved in something if it's gonna make me feel guilty down the road."

"Why would you feel guilty?"

"I don't know. On one hand, I feel like I'm the other person."

"And on the other?"

"Ah, I kinda feel like I'm cheating. I don't think I'm too excited about having an affair."

"Have you made a commitment to the girl back home?"

"No, ah, not yet at least. We're both free to do pretty much whatever we want."

"Well, then, what's the problem?"

"Theoretically, you're still engaged. When you told me, I should have walked away right then."

"Remember last Tuesday when I told you I wanted you?"

"How could I forget? Ah, as I remember, I told you the same thing."

"Well, did you mean it? I know I did."

"I suppose at the time, I did."

"I'm really glad I decided to take the chance to meet you. I've been attracted to you from the very first moment we met."

"I feel the same way, but it's still not right."

"Tell me you don't want to make love to me."

"I can't."

He shook is head, looking away.

"As long as you can live with it, I suppose I can, too, but I don't want you expecting something from me I'm not willing to give."

"Like what?"

"Ah, try empty promises. It would be easy to string you along, but the fact is, I'm in love with the lady back home."

"So, I'm in love with Vinnie. I don't see why we can't still enjoy each other."

"No strings, right?"

"No strings. Well, maybe there's one. I would want you to be here every night of the week. Why else do you think I asked you to move some of your clothing in?"

"Still not a good idea."

"Which, spending every night here or moving your clothes in?"

"Both. We're going to be shipping out for a cruise within the next couple of months. There's gonna be times when things come up I have no control over, and I'll have to stay aboard the ship."

Her eyes fixed on his, a seductive grin spreading across her face.

"Well, I guess I'll just have to make coming here every night so appealing you won't want to stay away. Admit it, Lee. Right now, you want me, don't you?"

His heart fluttered. Suddenly his stomach started churning. A surge of excitement invaded his groin and increased dryness in his throat discouraged speech.

"Yeah. I have since we first met."

"Well, then, why don't you put on a couple of records while I clean up the dishes? I'll be with you in a few moments."

He made a couple of hurried selections, and then leaned through the porthole to the kitchen to study her while she busily tended to the dinner dishes. Her white blouse was tucked neatly into a pleated green skirt, her trim waist flowing smoothly into the appealing curvature her alluring hips formed. Focusing on her upper torso, he noted her full breasts projected unobtrusively, yet stood out with clear definition. Reveling in her presence while attempting to control the excitement he felt spreading throughout his body, he closed his eyes and took a deep breath.

"About done?"

"Just finished."

She turned off the kitchen light. For a moment, she stood there eyeing him.

"Let's go into the bedroom. The couch gets a little cramped."

Shadows from the flames of the candles on the nightstand ebbed and flowed as he gently cradled the glowing body snuggled close to him, tenderly kissing her breasts.

Gasping, she moaned, "Oh, God, I could get used to this. You don't have to go back to the ship tonight, do you?"

"I don't have to be on board until a little before six in morning."

"Good."

A contented smile spread across her face.

"I'd like to visit the gates of heaven again."

"It's nice to want. Ah, I need a break."

"What's the matter? Did I wear you out?"

She paused, slowly running her soft, silky fingers up and down his chest.

"Ah, about not being able to visit every night, why? What could possibly prevent you from coming to visit every night?"

"I don't think you realize how unpredictable my schedule is going to be until we pull out. I have personnel records to keep current in case someone aboard our ship gets transferred. I also have to receive the people transferring to the ship. I'm also told we'll be going on maneuvers to test our engines and navigation equipment the closer we get to shipping out. As you can hopefully see, it's not just a simple as getting off work and going to the subway station so I can come to see you."

"I knew there was a reason I didn't like the Navy."

"Well, like I said, there's nothing I can do about it. It's probably none of my business, but what is there about the Mafia guy, ah, what keeps you interested?"

"I love him. You and he are a lot alike, packaged excitement."

She grinned, slowly raising her head until she could cradle it in the palm of her hand.

"The biggest difference is when he makes love to me. As soon as he fires his missiles, he's done. You don't seem to know quit, and it absolutely drives me crazy."

"If you're hinting, I still need a little more time to recover. Hitting singles and double isn't too difficult, but when you try to stretch them into triples or home runs, well, let's just say I need a little more time. About Georgan, I guess from what I can piece together, you were also involved with him."

"We were together until he found out about Vinnie."

She leaned forward and kissed him.

"I like the way you kiss. Um, do you feel like you're cheating on the girl back home still?"

"Yes and know. Ah, it's hard to know for sure the way you keep distracting me."

"Want to be distracted some more?"

"I'm getting there. Tell me, what's it like to go out with someone from the mob."

She grinned, sliding her hand down to greet his eagerness.

"There's something indescribable about getting fucked in the back seat of a limo while the a bodyguard stands watch outside."

He gasped, "I'll bet."

"Vinny isn't gentle like you are. He's hung like a horse, and, God, when he rams it in, I feel like someone's just driven a truck inside me. You're not huge like he is, but I'll give you a ten for style points."

"You're not so bad yourself."

"Oh, my!" Chuckling seductively, she solicited, "I think you're just about ready to go again. Mind if I tell you something, first?"

"Go for it."

"You're going to need patience if we're gonna get along."

"How so?"

"I have a temper, and, ah, it gets a little nasty sometimes. When I go off and my temper gets the best of me, just remember it's the Scottish blood taking over. I don't mean anything by it. I'm just venting."

"I'm supposed to ignore your behavior and assume you're just having one of your temperamental Scottish fits, right?"

"Yep. If it's anything more serious, I'll let you know."

He smiled, gently caressing her breast.

"I think when I see one of your fits coming, I'll just step back, and hope I don't get hit by flying objects."

"Good plan."

"Damn, this whole scene kinda blows my mind."

"Why?"

"Most girls want to be in love before they give it up."

"I'm not like most girls. I enjoy fucking, and I'm not afraid to admit it, but I'd like to think you're mine every night."

"Ah, you mean every night except for the weekends."

Father Time's apparent impatience made the approaching deployment of the *Tanner* seem to arrive more rapidly than expected. As the June deployment neared, the approaching reality Lee would be gone for and extended period started to increasingly become an issue between them. As the hectic preparations for the mission to the Barents Sea escalated, an increased workload in the Ship's Office started to delay with increased frequency, even interrupt his visits. Tension was starting to build between them.

This involvement with Peggy is starting to get out of hand. She has an insatiable appetite. Fuck, the only time I get a break is when her time of the month comes around.

He sighed audibly, frowning with a mild display of displeasure.

Peggy's a great gal, but I'm in love with Sandy. So, if she's the one I plan to spend the rest of my life with, what am I doing messing around with her all the time?

Things have never been better with Sandy. I think when I go home on leave, I'm gonna knock on her door. Yeah, in the meantime I've got to end this what ever you call it with Peggy. It's not fair to string her along. The sex has been great, and all the things we've been able to see and do with each other have been fantastic, but it's still not enough.

He smiled recalling their strolls through Central Park, watching the young yachtsmen sailing their boats on Central Park's lake, and quiet moments spent together atop the Chrysler Building watching the panorama of the city lights.

Okay, visiting the Rockefeller Center, catching a movie at Radio City Music Hall, and taking the ferry to see the skyline have all been great, but something's missing. Five minutes on the phone with Sandy is so much more meaningful than all the time I spend holed up with Peggy. Yeah, I've got to end it before I mess things up with Sandy.

"Do you realize we only have two more Monday's together before your ship pulls out?"

She shook her head as she put the last dish in the cupboard.

"I'm really dreading it. Ah, looks like the situation with Vinnie isn't gonna work out."

"Oh, what seems to be the problem?"

"The family Vinnie works for is under investigation. I guess the old man wants Vinnie to stay pretty close to the daughter now the Feds are trying to get her to turn state's evidence. Maybe the old man has a reason to fear his darling daughter will spill the beans on the family's operation."

She frowned, a detached haze enveloping her gaze.

"There's even talk that he'll have to relocate in Seattle. Hmm, maybe Vinnie's been stringing me along all this time. Are you still keeping the airways and postal lanes busy with the girl back home?"

"I've never lied to you, Peggy. I told you I love her."

"You told me you love me, too."

"Yes, I did. It's just, ah, well, I love you in a different way."

"I'm just a passing fancy, someone to fuck, huh?"

"No, it's much more. You're a good friend."

"Not good enough to consider marrying, though?"

"No. Look, we both went into this with our eyes wide open. Neither one of us has hidden anything from each other. Lately, I've been thinking we should stop seeing each other. It might make my leaving a little easier."

"Sometimes you come up with the dumbest ideas. Ah, you've only got a couple of weeks before you pull out. Why quit now?"

"Oh, I don't know. Maybe I don't want you to think I'm using you."

"Well, aren't you?"

"I'd like to think I wasn't."

"What if I told you I was pregnant?"

Lee's face paled, tears welling up in his eyes. From the deepest reaches of his throat to the pit of his stomach, a searing, burning sensation was ignited.

"Fuck! Please tell me you're kidding."

"In case you don't know it, you've been treading on pretty risky ground. If I weren't taking birth control pills, I'm pretty sure we'd be looking parenthood squarely in the eye."

"Whew. You really had me worried for a moment. Ah, it's not that I wouldn't have…. Ah, let's just say…."

"Yeah, you're happy as hell I'm not. So, you want to keep seeing me until you ship out for the Barents Sea?"

"You'll have to realize the next couple of weeks are gonna be even more unpredictable than ever. I probably won't see you every night."

He rose from the couch.

"I have a big day tomorrow. If I stay here tonight, I won't get much sleep, so, I think I'd better go."

"Will you be able to stay tomorrow?"

"Maybe, but it won't be until about eight. Ah don't bother saving dinner."

"Oh, ho, want to get right down to business, huh?"

Moments later, standing on the subway platform waiting for his train, an internal debate was raging.

I should have ended it, tonight. I'm such a chicken. I also should have told her we were leaving next Monday and about the sea trials tomorrow. Every time I'm about to end it, she flaunts that fantastic body and that irresistible charm of hers.

He took a seat on the subway bound for Brooklyn.

I feel like a real heel, but I'm in love with Sandy. She seemed to be really excited when I told her I was planning to knock on her door when I came home on leave. Yeah, next Monday will be the last time I see Peggy. I just hope it doesn't get ugly.

Outside Peggy's office, the following Monday, he smiled as she reacted in shocked amazement.

"Um, the reason I haven't seen you before this is, ah, we've been on maneuvers. Ah, we're shipping out early."

Peggy frowned, placing her hands on her hips defiantly.

"How nice! When's the big day?"

"Tomorrow. I have until noon tomorrow."

"You say you've been on maneuvers. Are you sure the girl back home isn't the reason you've been staying away?"

"I'm not going there, again."

"So, where's your uniform? You're not gonna say goodbye to me just like you said hello?"

"I changed at the YMCA."

"So, you must be planning to stay all night."

"Possibly. It doesn't hurt to be prepared."

"I've missed you. I sure hope I can convince you to stay."

"I could be persuaded. What would you like to do?"

Peggy grinned sarcastically.

"How novel. I'm starting to wonder if I shouldn't join the convent and you ask me what I want to do."

Suddenly with warning, she exploded. It wasn't a quiet Fourth of July sparkler either. More like the rocket explosives mushrooming in the air, it was visible and audible to all.

"I get so dammed tired of the fucking Navy I could vomit. If I didn't know better I'd think you were having an affair."

He blushed, not a pale hue, either. It was a full blown, bright red floral display.

"Shhh."

"Don't shhh me. What makes you think you can just drop in to see me whenever it's convenient? I can't even count the number of dates you've broken the past few weeks."

People passing by the noisy outburst started to stop and stare. Frustrated, Lee thrust his hands into his pockets and stared at the pavement.

"Please, let's not argue, not here. Can't we at least wait until we get to your place so we can discuss this matter without sharing our story with the entire population of Manhattan?"

"What's the matter, bashful?" Sneering, she shouted, "I thought you would take great pleasure in announcing to the world you were fucking me."

She turned to the gathering crowd.

"Look at me, everybody. My lover has a mistress. She's long and gray. She has numbers painted on her chest, and she swims like a fucking eel. When this guy gets bored staying aboard his fucking ship, he sneaks off and tries to nail me."

Lee chuckled nervously as an older lady stopped to stare at them.

"I can't help it if I have obligations."

"Ha! You don't have obligations. You have a fucking obsession."

Somewhere from the background of the crowd, someone shouted, "For Christ's sake man, move it along. Your performance is holding up traffic. Besides, I could see a better fight at the Garden."

Later, Lee inhaled deeply after recovering from the frantic intensity of the exertion.

"Feel better?"

"And how! I love you, Lee Grady. I just wish I knew from one minute to the next whether to expect you or not."

"So does the crowd we attracted."

He leaned forward and gently embraced her.

"This is the last time we'll be together. Ah, it could be for a long time."

"Yeah, and then again, it could be forever."

He playfully patted her soft rump.

"You were right. When you start to vent, you're like an erupting volcano."

"I warned you, remember?"

"I remember. I just didn't think…."

Nodding, she pleaded, "I just wish, ah, yeah, I know. We didn't make any promises."

"No we didn't make any promises."

Early the next morning, playfully playing a tune on the keyboard of her backbone, he probed, "So, what would you like to do the rest of the morning?"

"Oh, I don't know."

"What else do you have in mind?"

She got up from the bed and slipped into the robe lying at the foot of the bed.

"I'm going to jump in the shower. Um, a friend of mine is dropping by to look the apartment over. If she gets here while I'm in the shower, will you please show her around the place?"

"Why is your friend looking the apartment over."

"I'm moving back to Connecticut."

"Hmm, when did you plan to tell me?"

"What difference would it make? Ah, I got another job while I was home this past weekend. I gave notice Wednesday, so I guess next week will be my Bon Voyage to the Big Apple. Um, I probably should have told you before, but with you gone all the time, I never had the chance."

She glared at him, a wicked sneer enveloping her expression.

"Vinnie and I are going to move in together."

"What about his wife?"

"She's gonna stay in Boston.

"Hmm, I guess that means you'll be a kept lady, of sorts."

"At least I can count on him showing up. Ah, you'd better hurry and get dressed. My friend will be here any moment, and, I'm not real sure she'll enjoy seeing you in the buff."

The doorbell interrupted further discussion. Quickly pulling on his pants, he answered the door.

"I'm standing in for Margaret. Ah, she's in the shower."

"My, my, look at what I've found."

A petite brunette impressively occupied the entryway. Grinned slyly, she entered.

"You're Lee, aren't you?"

"Yeah, and you must be Bobbie."

"I've heard all about you. If you don't mind, I'll just browse around."

"Yeah, ah, I mean no. Go on ahead. I was supposed to show you around, but you'll probably do better without me getting in your way."

"Darn, I had my hopes up for a moment. Ah, don't worry about my friendship with Peggy, honey, we're not lovers."

"Refreshing! Since you're not lovers, you might want to skip the bathroom tour until she's done."

"Right! I don't need to see the bathroom, anyway."

Quickly she looked around the living room and kitchen. After a quick glance into the bedroom, she shrugged.

"All done."

"So soon?"

"I'm quick. Ah, tell Peggy I'll take it. Tell her I'll call before she leaves for Connecticut. By the way, she talks about you all the time. She didn't lie, but I'm just not into the guy thing."

"Yeah! It was a pleasure to meet you, Bobbie. I'm sure Peggy will be pleased to hear you want the flat."

A couple of minutes later Peggy emerged from the bathroom, toweling her wet hair.

"Has Bobbie come by yet?"

"Yeah, she'll take the flat. She said she'd call you."

"She's gay, you know."

"I figured as much when she mentioned she wasn't into the guy thing. The way she looked me over though, you think she mistook me for one of her lady friends. Ah, I'm just about ready to take off. I've got some things to do before we set sail. Wanta see me off?"

"If you think I'm going to stand on some fucking dock with the rest of the fleet honeys, think again. Ah, before you go, I want to give you a letter I've written. I'd appreciate it if you'd wait to read it, though."

"No problem. Um, it's been good, Peggy."

"We just didn't have enough going, huh?"

"No. Ah, I've decided to ask Sandy to marry me."

"Well, good for you. Um, with Vinnie and I getting together, it wouldn't work out anyway. Ah, in the letter, I gave you my new phone number and address. You can write me while you're on the cruise if you'd like. Who knows, when you get back, maybe we'll both feel differently."

She paused, tears starting to trickle from the corners of her eyes.

"How does it go, absence makes the heart grow fonder?"

"Sometimes."

At the subway station, he sighed, relieved that their parting had gone so smoothly.

Hmm, ten o'clock. I'll still have an hour after I get changed before I have to be back on board.

He reached into his pocket and took out the letter Peggy had given him. Opening it, he quickly read the contents. Nodding, he held the note over the grated drain next to the street curb. Slowly he released it and watched it flutter and dart with the currents of the mild early summer's breeze. When it reached the grate, it hovered on the lip of a narrow metal rafter, and then slowly slipped into the dark grave. He shrugged as he peered into the tomb for a glimpse of the paper.

Hmm, as it began, so it ends.

THE JOURNEY

Often a person's life is related to an environmental phenomenon such as the wind, a mountain, a river, or even some animal. "My life is like a river," expresses a search for commonality between a person's life and the watery confluence. Noting the path the river follows, it is not unlike observing the trail we follow passing from one adventure to another. The river, a marvelous watery organ, moves and surges along through its environment displaying the many sides of its personality. A bend here or a determined forward movement there dictates how the stream can change from a peaceful, lazy traveler to a violent, energetic excavator. Whatever its mood, it makes its way onward to seek a final destination, moving in and out of many different influences, finally spilling into the sea or some other body of water. There its identity merges with the other essence. The river, like man, experiences the seasons of life. Geologically it is born, grows or blossoms, and matures, aging until finally dying. Every river has a story. Although it does not appear to have the language or the knowledge of human expression, it finds remarkable ways to relate its message.

The *Tanner* returned to the Brooklyn Naval Yard in early December from a nightmarish three-week cruise to the Bahamas. It was during the unexpected adjunct to the Barents Sea mission, the day before Lee's birthday that the nation lost its thirty-fifth president.

Two weeks later, Lee's parents greeted him at the United Airlines terminal.

"When you called, you said you'd be able to be home until the middle of January."

"That's the way it looked at the time. Um, I'm going to be spending a lot of my time with Sandy. I hope you guys understand."

Lee's dad smiled, nodding his head in approval.

"Glad to hear it. Ah, you might want to know she just won a big case. The rumor has it she's pretty hot property right now in the criminal justice system."

"Yeah, I guess she's making a career move. Have you heard anything?"

"Rumor has it she has been offered a position in U.S. attorney's office in Portland. Didn't she tell you?"

"No. I guess she's saving the news as some sort of surprise."

"Shouldn't pose a problem. Ah, Portland's only forty minutes from Salem.'"

His dad smiled slyly, continuing his quest for his son's status with Sandy.

"So, what's up between you two? Is it just wistful thinking, or have you two been talking about something a little more than maintaining a friendship?"

Lee's face reddened.

"The topic has come up."

"And?"

He reached into his pocket.

"Um, I bought this ring just before I left New York. It's not what a person in her position would expect, but it's all I could afford."

"If she loves you, it won't matter. Just be damned sure you're certain. We're talking about the rest of your life here, son."

"I know. All the time I spent in the Barents Sea helped me sort a lot of things out. I just hope she feels the same way I do."

The next evening, he rang the doorbell of Sandy's South Salem home. A moment later she opened the door, a big smile consuming her face.

"Well, are you going to just stand there staring at me, or do I get a hug?"

More beautiful than ever, the casual evening dress she wore accented her many qualities. He grinned.

"I was just letting my eyes get used to the bright light cast by the most beautiful women I've ever known."

Stepping forward, he and took her in his arms, and tenderly kissed her.

"It's been too long."

"I've missed you, too. No, it's more than that. I've been counting the minutes since I last saw you."

He grinned.

"You've been busy, then."

"Yeah, Um, the place we're going to this evening, is a favorite of mine. I was thinking a little romantic retreat might be in order. What do you think?"

"Perfect! I hope this place isn't offended by casual dress. Am I okay?"

She stepped back, surveying him carefully. She smiled, nodding.

"Your sports coat and slacks look nice. Um, I forgot you hated wearing a tie, and unfortunately tie is considered a necessity if you plan to dine at Henry Ford's."

"You're right, I hate wearing a tie with a purple passion, but I did come prepared." Motioning towards his car, he disclosed, "Um, I left it in the car."

"Good, you can put it on while we're driving up to Portland. Give me a minute to get my coat."

A moment later she returned, chuckling as she locked the door.

"You and your love for ties amazes me." Taking a step forward towards where his car was parked, she suddenly stopped.

"Um, on second thought, go get your tie. I think I'd like to take my new car."

"You got a new car? Wow, what did you get?"

"What would you say if I told you I bought a new Corvette?"

"Are you kidding? You really bought a Corvette? Damn, I'm so jealous."

"You've mentioned so often how much you wanted one, I guess you finally won me over."

"You're can't be serious. It's really new?"

"It's blue with white trim, just like your eyes. Hurry up and get your tie so we can go take a look at her. I parked it in the garage so I wouldn't spoil the surprise. Wanta drive?"

On the I-5 freeway, he nodded appreciatively.

"I'm starting to get the feel of this baby, and let met tell ya, this puppy is flat out hot."

He grinned, glancing at her.

"Have you opened her up yet? I'm feeling power I'm almost afraid to tap."

"Go ahead and punch it." Grinning, she confessed, "I've put the pedal to the floor once or twice. Um, this freeway is a good place to see what she can do. Go on, punch it."

"Nah, I think I'll pass. Ah, maybe some other time when the traffic's a little lighter and I get to know this baby a little better. Besides, it'd just be my luck to get nailed by some cop."

She reached over and put her hand on his.

"Other than talking about my car, you haven't had much to say. Are you trying to decide if you're just dropping in for a visit, or do you plan to knock on my door and stay awhile?"

"Let's say, just for argument's sake, that I was planning to knock. Would you answer?"

"Why don't you try and see?"

"I don't want to play games, Sandy. I know your career is established and moving forward very rapidly. Hell, I'm still in the starting blocks. Ah, I don't have shit to offer."

"You have more to offer than you realize."

"Sure I do. Um, you mentioned a career change. Is the rumor true about you moving to Portland true?"

"I accepted a position with the U.S. attorney's office. I start January 20th, the day after your leave is up."

"So, I guess you plan to sell your home and move to Portland?"

"No. For a while I'm going to commute. I really like my home, and, besides, I haven't had much time to look for housing in Portland. Um, I was thinking I'd like to talk with you about the living arrangements before I made any final decisions."

"Is that a subtle way of saying you'd answer if I knocked?"

She smiled, pretending not to hear him.

"Um, where we want to get off the freeway is the coming up. Our exit is the one after the 217 exit."

A few moments later Lee wheeled the blue and white Corvette into the parking lot. A parking attendant took his keys and directed them towards the doorman.

"I'll keep these while you and the lady are dining. When you're ready to leave just tell your waiter Josh is your parking attendant. Your car will be waiting for you at the entrance."

Inside, the maitre d' showed them to a table in the corner of the dimly lighted dining area. Once they were seated, he expertly acquainted them with the menu offerings.

"Your drinks, rum and Cokes, have been ordered. They will be here shortly."

Sandy smiled at the waiter, nodding knowingly.

"You did take care of everything as I requested, didn't you Morgan?"

"Yes, Miss James, everything is all set."

Winking, she turned to Lee.

"Honey, I forgot to tell you that tonight's my treat. Now, let's get back to our discussion. I still want to know if you plan to knock at my door."

"Back on the freeway, I asked if you would answer. I'm still waiting for an answer."

Still awed, he shook his head.

"Can you believe our waiter. My God, he rattled off the list of appetizers, the entire selection of entrées without even using a menu. How did he know what drinks we would order, let alone when to have them delivered to our table?"

Sandy smiled slyly.

"Oh, I don't know. Ah, he's good isn't he?"

"You don't do him justice. He's an ace. I'll bet with his talent, he makes big bucks."

He smiled, leaning forward towards her, gently he taking her hand and looking into her eyes.

"I still want to know if you'll answer."

"Let's say the answer is yes, what do you have in mind?"

"How does a full partnership sound?"

She squeezed his hand affectionately as their eyes met and fixed upon each other.

"I love you, Lee. I've loved you since the first moment we met."

"I've had a lot of time to think everything through. Being at sea gives a person a lot of free time for thought. Ah, I've considered my future, and, in particular, a possible future with you. It has taken me awhile, but I know exactly what I want."

"What are you trying to say?"

"I want to grow old with you. I've met a lot of people and shared many experiences, as I'm sure you have, but I'm tired of meeting new people and having new experiences. I met the person who pleases me a long, long time ago."

"And? A girl can only stand so much suspense." Motioning towards a far corner of the room, she disclosed, "The man with the violin is only going to stand there so long before he loses interest."

He blushed, slowly withdrawing a small box from his coat pocket.

"I have something for you."

"Lee?"

Tears appeared in her eyes as she gently cradled the white box.

"Is this what I think it is?" Slowly opening the box, she squealed, "Oh, my God!"

The small diamond stone reflected a glint of spectral light from one of the candles on their table, its luster sparkling and bright like Sandy's eyes.

"It's so beautiful. Would you put it on my finger for me? I'm too nervous."

"I'm a little nervous myself."

"So, is this your sneaky way of proposing?"

He grinned, gently slipping the ring on her finger.

"I could ask about the violinist."

Blushing, she implored, "You could, but I wouldn't say a word until you give me an answer."

"Then my answer is yes. I decided before I bought this at Tiffany's that I wanted to knock at your door. Will you honor me by answering?"

"I thought you would never ask."

Her eyes started to mist as he gently held her hand and gazed into her eyes.

"Of course I'll marry you. It seems like I've waited forever for this moment to come."

He rose, leaning across the table to greet her lips, the violinist beginning the serenade. After a moment their lips parted.

Chuckling, he whispered, "Did you orchestrate the applause, too?"

"No, silly. I'm only responsible for the violinist. Everything else is totally spontaneous."

"So what would you have done if I hadn't asked you?"

"Are you kidding. I wasn't going to let you get away this time. In case you haven't figured it out by now, I love you, and I've planned on sharing my life with you for a long, long time."

SECOND CHANCES

As with a river, a turn here, a bend there, suddenly three decades have passed. Miles of winding travel have passed, many challenges met with rewards too many to list. In the quiet latter seasons of life, time for consideration and assessment has finally come.

On a clear, warm Friday afternoon in 1998, Lee busied himself cleaning the driver's inside window of his red Ford pickup.

One more window to go and I'll have this beast all cleaned up.

He cocked his head, straining to detect the source of an unexpected sound.

Sounds like the phone's ringing. Damn, I hate telephones! Every time you start to do something, they interrupt.

He grunted, reluctantly lifting his tired body from the bucket seat on the driver's side. Hesitantly, he moved in the direction of the portable phone located on the freezer. It rang again.

No need to hurry. Whoever it is will probably hang up before I answer, anyway. On second thought, I think I'll go in the house and listen to the answering machine to see whom it is. Probably some damn solicitor, anyway.

The answering machine kicked in with the bland announcement he had recorded just as he reached the kitchen.

"You've reached the residence of Lee Grady. By now you've discovered

I'm gone or unable to answer the phone. If you're so inclined, try reaching me later, or wait until I get back to you. Solicitors might be interested to know they're having a memorial service next door in my honor. Anyone else can leave your name and the number where you can be reached."

"Dad, this is Nicole. Charon and I have something we want to show you. When you get home, give me a call."

Quickly he reached for the phone.

"Don't hang up, honey, I'm here. I was out washing the truck. What's up?"

"If you're gonna be home, Charon and I would like to drop by. Ah, we have something to show you."

"Sure, come on over. I'll be done with the pickup by then. Is your hubby coming, too?"

"No, Jason had to go into the school for a few minutes. Ah, see you in a few."

An hour later, Lee greeted his two daughters.

"You said you have something to show me."

Charon nodded, appearing to be surprisingly mysterious.

"We brought a letter we've been holding. The time has come to show it to you. Um, Mom gave it to us before she...."

"Really! Is this some sort of conspiracy?"

"No, ah, it's just something she wanted us to hang onto until it was time to show you."

Charon looked to Nicole for support.

"Dad, the letter has to do with the Anatolia Mafia family Mom prosecuted years ago."

"The case was over before either of you were born. I don't know why she'd write me a letter about the case. The head of the family died in prison, years ago. Yeah, even his top lieutenant, Vinnie Trabago is gone."

Nicole nodded.

"Killed in some sort of prison revolt, wasn't he?"

"So I heard. Hmm, I wonder what's so important that Sandy would feel compelled to put it in a letter and not share it directly with me?" Frowning, he probed, "And why did she tell you girls, instead?"

Charon handed the letter to her father.

"Here, read it. Everything is explained in the letter."

"Have you read it?"

"No, Mom just told us what was in the letter. She told us just before…."

"Strange. Um, would you like for me to read it aloud?"

Tears started to fill Nicole's eyes.

"Sure."

Tearing at the sealed envelope, Lee was interrupted again by the unwelcome intrusion of the telephone.

"Damn telephone. It rings at the most inconvenient times. Ah, screw it. Just let it ring. If the caller wants to talk with me badly enough, they'll either call back or leave a message. Come on, let's go into the living room and take a load off while I read this thing."

Taking his usual seat on the couch, Lee took the letter out of its enclosure.

"Hmm, scented and everything. Well, here goes. Let's see what your mother had to say. Ah, she began, 'My Dearest Lee.'"

He looked at his daughters and then back at the letter.

"Yeah, well, she wrote, 'You will receive this letter long after the case against T. Anatolia and his son-in-law, V. Trabago, and the entire Anatolia Mafia family has been resolved. I have, as requested by Pat Brown, asked Nicole and Charon to give it to you on the one-year anniversary of my death.'"

A tear appeared in the corner of his eye. Uncomfortably, he cleared his throat.

"Ah, she went on to say, 'As you know, irrefutable evidence led to the conviction of those despicable criminals, resulting in a life sentences in a federal prison for both of them. The investigation leading to their prosecution and the rest of the Anatolia Mafia family consumed an enormous amount of my time, dating back to 1962. In building the case, brick by brick, my prosecuting team and I were able to bring charges and prove T. Anatolia, V. Trabago, and other key Anatolia family members were guilty of money laundering, income tax evasion, drug trafficking, jury tampering, bribery, murder, and many other crimes. I regret I was unable to share all the details of the case with you, but for two very important reasons, I was forced to remain silent.'"

Lee frowned, his furrowed brow indicating that the wheels were turning.

"Hmm, all the names of those involved, ah…. Anyway, your mother's letter goes on to say, 'An interesting triangle existed involving the defendant, T. Anatolia, also known as Trish Knight, Vinnie Trabago, and the prosecution's key witness, Pat Brown. When you came home on leave in 1962, and you told me about the death of Trish Knight, If you'll recall, I asked you about her family, inadvertently mentioning Anatolia Enterprises. At the time I had just started investigating some suspected criminal activity of the Anatolia Mafia family, but I had no idea I would uncover a hideous cover-up plot, directly involving you.

'As you know, Pat Brown is the name of our key witness. She assumed the name when we placed her in the witness protection program in exchange for her testimony. She was also known a Patti Anatolia, Patricia Anatolia-Trabago, and Trish Knight. In an attempt to protect a beautiful memory and because of her insistence, I never mentioned the specifics of the case or revealed the true identity of Pat Brown.' Well I'll be damned."

He paused for a moment, reflecting before again beginning to speak. Slowly, he nodded.

"Now everything is starting to fit. Humph, I guess everyone knew what was going on except me."

Lee fingered the letter tentatively, another tear appearing in the corner of his eye.

"She worked on the damn case for a long time, too long. Ah, I never asked her about any of her cases, but the Anatolia case, ah, well, it always seemed kinda funny to me."

"What do you mean, Dad?"

"I don't know exactly. I always had the feeling Sandy knew something she was afraid to or was unwilling to share with me. Ah, I've always felt I was connected in some way. Now I know."

"Mom couldn't say anything, Dad, she was sworn to secrecy."

"I know, yeah, now I know. Ah, but why did you keep it quiet all this time?"

Bristling, Charon defensively snapped, "Because we couldn't. You surely wouldn't have expected us to say anything, especially since Mom asked us not to."

Nodding, he conceded, "No, I suppose you couldn't." Smiling, he quickly added, "And I proud of you for not telling me, ah, until the time was right."

"Why don't you finish reading the letter, Dad. We can talk about everything when you're done."

"You're right. I'm sorry I jumped on you. I'm just sort of reeling right about now. Well, anyway, the letter goes on, 'You are probably wondering why and how you are related to the case. Trish was afraid if you found out she was still alive, you would try to contact her. Her father, Trush Knight, had insisted she not marry you, and even went so far as to threaten to have you killed if you even so much as made contact with her. It's the reason she agreed to participate in the faked death hoax. She also told me her father had tried to force her to have an abortion. By agreeing to his plan, she was able to save her baby and your life. To illustrate what an evil man he was, you should know he threatened to take her son away from her or do harm to him on a regular basis. It was all in attempt to keep her in line after our investigation started to heat up in 1964. I guess she got tired of living in fear because just after you came home on leave in 1964, she made contact with my office through a police officer in Boston. When I met with her, she agreed to testify and go into the witness protection plan when I assured her we could guarantee the safety of her son. We also promised we could insure your safety.

"'Trish is quite a lady. When she found out you and I were getting married, she upped the ante, agreeing to testify only after receiving an assurance we would never divulge her identity. After I got sick, I discovered I only had months to live. I made contact with Trish because I wanted to be able to set the record straight and let you know the truth. I also wanted you to be able to meet and get to know your son. Trish agreed, but thought it best to not break the news until you had time to adjust to my passing.

"'I am sorry to have been forced to keep this a secret for so long. If I had been able to, I would have told you years ago. I just hope you can understand and forgive me. I didn't want to deceive you, nor did I want to involve Nicole and Charon in the drama.

"'I know I have no right to ask, but please listen to what the girls have to say. They love you so much and only want what is best for you. I'll love you forever, and remember we have an appointment in Heaven.' Wow! Well, at long last the mystery is solved."

His eyes filled with tears, as the letter slipped from his fingers, falling into is lap. Uncomfortable with the outward display of emotion, he quickly wiped his eyes.

"Oh, yeah. God, I hate it when I start to lose it." Smiling sheepishly, he revealed, "Ah, she signed it, 'Love you forever, Sandy.'"

Charon got up and started for the kitchen, a devilish smile etched on her face.

"I'm going to check the message machine to see who called."

"Why? Whoever it was can wait."

"Because, I'm expecting someone to give you a call."

"Well, have it your way, but hurry up and get back in here. I have some questions I want to ask you and your sister."

A moment later, Charon yelled at him from the kitchen.

"Dad, listen to this message. I'll turn up the machine so you can hear it."

The recorder started to play.

"This is Pat Brown, Lee. Please give me a call at (503) 228-2000. Ask for room 324. I'll be at this number all evening. It's Friday and the time is approximately two thirty."

"Is she the caller you were expecting?"

"Yeah, I called her, Dad, and told her it was time to meet with you to set the record straight. Ah, Nicole had nothing to do with this."

"Well, I guess if I'm gonna put this baby to rest, I'd best give her a call."

"Promise you'll hear her out?"

"Yes, Dad, give her a call." Charon paused in the doorway leading to the kitchen as tears started to cause her eyes to glaze over.

"Dad, Mom loved you so much. The last thing she asked us to do was make sure you found out the truth. Now, go make the call and allow yourself to catch up on thirty-six years of history." Wiping her eyes, she added, "It's long overdue. Besides, wouldn't you like to meet your son?"

"What about you girls, you're okay with this?"

Nicole nodded.

"Are you kidding? Of course we're okay with it. Um, I think we're going to leave. This is something you need to do by yourself. You don't need us here eavesdropping."

Charon nodded, a devilish grin spreading across her face.

"You could call us later and let us know all the details, if you want to." Suddenly overcome by emotion, she choked back a tear, "Mom would have wanted it this way."

"Okay. I'll make the call."

He shook his head, a large smile starting to sweep across his face.

"It's probably the only way I'm gonna get you off my back. Now go on and get out of here. I have to take a shower and get ready for my dinner date with Mel and Lynne."

"Don't forget to make the call today."

After his girls had left, he started to get ready for the evening engagement.

Hmm, three o'clock. Mel and Lynne won't be here to pick me up for dinner until a little after six. I guess I'll get ready and then catch the Mariner game.

Returning from the garage where he had said goodbye to his daughters, he made a beeline for the telephone.

I might just as well call her back and get it over with. It won't hurt to find out what she wants before I crawl into the shower. Hmm, Pat Brown, Patti Anatolia, Patricia Anatolia-Trabago, Trish Knight, I wonder why she wants me to call her? Funny how all those names.... Ah, all except the Pat Brown handle have been popping up since I first found out about Trish's accident.

"Benson Hotel, how may I help you?"

"Connect me with Pat Brown's room, please. Ah, she's in room 324."

"Certainly, just one moment please."

A moment later, a distantly familiar voice replied, "Hello?"

"This is Lee Grady. You asked me to call."

Her voice, mature and pleasant, contained an almost hopeful tone.

"Thank you. By now, I suppose you've read Sandy's letter?"

"Yeah, ah, all of this has hit me like a ton of bricks. I'm just starting to get past Sandy's passing and now this."

"I know. I'm sorry it had to work out this way. Ah, the telephone is a lousy way to catch up. I was hoping we could have dinner or go out for a drink this evening."

"I don't mean to be rude, but I think I'd like to find out a few details, first. Ah, I suppose you know what was in the letter?"

"Yes, Sandy and I discussed what she was going to tell you when she told me about her illness."

"Good. Then why don't you fill in some of the gaps?"

"I feel like I owe you so much. Somehow this is like I'm trying to pay off a thirty-five-year-old debt, but I'd like to do it face to face."

"You're probably right. The telephone would be a lousy way to uncover this mystery, and, besides, I hate the telephone. Gosh, do you realize it's been over thirty years since I saw you last. Are you doing okay?"

"I'm fine, now that I'm able to finally talk to you. By the way, to be exact, it has been a little over thirty-five years, much too long and so unnecessary. About dinner, are you free?"

"I have plans for this evening. I'm going out to dinner with some friends." He paused, the phone line suddenly rendered lifeless.

"Ah, if you don't object, you're welcome to join us."

"You don't mind?"

"Not at all. I'm sure my friends will understand."

"Good. Ah, I'm not pressed for time, so maybe we could talk after dinner."

"Kinda what I had in mind. Ah, this is gonna be awkward. Mel thinks you're dead, and I don't know a damned thing about you. Shit, I don't even know who you are anymore. You're gonna have to give me a little help, Trish, Pat, or whatever. Um, it is okay to call you Trish, isn't it?"

"Trish is the only name you've known me as. As for any specific details about me, if you just tell everybody what was in the letter, you'll be fine. Anything else is just filler, something probably best kept between you and me."

"Um, is your mother still alive?"

"Yes, she's still living in Seattle. She asked about you, recently."

"Like you had a lot to tell her. Ah, how are things between you and her? You did have to testify against your father, you know."

"The relationship was strained for a long time. In fact, until just recently, we were totally estranged."

"What changed it?"

"When she started to sell off pieces of my father's empire, she needed my okay. I'm an heir, you know. Well, anyway, we got together and started talking. Little by little, we buried the past. She decided having a relationship with her grandchild and me was more important than holding a grudge. Lee, She's not like my father. She's not Sicilian."

"Well, good. I'm glad to hear everything is good between you and your mother."

"Um, Lee, just to set the record straight. You must know that I was never really married to Vinny. There was a wedding ceremony and we lived together for awhile in Boston before I decided to testify, but we never shared the same bed. I'll tell you more about the Vinnie experience later. Ah, do you suppose I've shared enough information for now? I'd kinda like to wait until we have more time and we can talk face to face before I share the rest.

"Can you be ready for dinner at six-forty-five? We have dinner reservations at seven."

"I'll be ready. Should I wait in the lobby?"

"Why don't you meet us out front on the Oak Street entrance? It will be easier. Finding a place to park in downtown Portland is such a pain in the ass."

"Sounds like a drive by pickup to me. Intriguing. I'll see you at the Oak Street entrance about six-forty-five."

"I'm not sure I'll recognize you."

"I'll probably be the only sixty-year-old blonde waiting on the street outside the Benson. I'll be wearing slacks and a green sweater. I hope I'll be dressed okay. I didn't ask about the dress code."

"You'll be fine. This is a casual evening."

After placing the receiver in its cradle, feelings of ambivalence started to intrude. Trish had stirred up old feelings. Teetering on the cusp of curiosity and exhilaration, he felt guilty for allowing himself to feel the way he did. He still missed Sandy terribly, but somehow, he couldn't explain the elation he felt about knowing he would soon be able to see Trish again.

It's gonna be nice to see her after all of these years, but it kinda pisses me off she's been alive all these years and this is the first time I hear about it.

He stared at the kitchen wall trying to figure out exactly how he felt.

Guess I'd better call Mel and Lynne and tell 'em we have a surprise guest.

"Hey, Mel, this is Lee. We're still on for six-thirty, aren't we?"

"Sure, unless you want us to get up there earlier."

"No, six-thirty is just fine. Um, what would you say if an old friend of mine joined us for dinner?"

"A lady friend?"

"Yeah, someone from my past. Ah, I'll tell you and Lynne more about it tonight. Right now I'm still recovering from the shock. Sure you don't mind?"

"Of course not. Lynne will be thrilled to have someone to talk to when we get started on another one of our fishing tales. Besides, I'm curious to meet this mysterious person from your past."

"Well, when I tell you the details, you'll understand what I'm going through. Ah, I'll see you about six-thirty."

Guess I'd best hit the shower. These old bones are really starting to bark. The hotter the shower is the better!

He turned on the water and adjusted the temperature.

Ah, just right. Boy, I'll tell ya, the news about Trish is unnerving. A long time ago, yeah, sure, I was in love with her, but now it's almost like it never happened.

He stared at the fine spray coming out of the showerhead.

I'm not sorry in the least the way everything turned out except for Sandy leaving me so soon. Nope, I wouldn't change what's happened in my life one iota.

He shook his head, wrinkling up his brow.

It's a bit scary to think how close I came to marrying Trish. Fuck, I would have missed out on getting together with Sandy, and Nicole and Charon would never have been born. Funny, at the time, I probably even wanted to marry her. Strange how things work out.

Slowly he started rub the bar of soap across his arms and shoulders.

I wonder if we were really in love, or what? Everything developed so fast. Maybe she was just rebelling against her father.

He studied the bar of soap, attempting to gather his thoughts.

Even though Charon called her to set all of this up, if Trish didn't want to honor Sandy's request, all she would have had to do is refuse to see me. There's no way she had to call.

He turned so the spray could hit his back.

In the letter from Sandy, she mentioned something about Trish's father having me killed if I found out about the hoax. I wonder why? I couldn't possibly have posed a threat to him.

He shrugged as he continued to laze under the hot spray, mechanically placing the bar of soap in the soap tray.

Ah, the pulsating action feels so good.

He lazily glanced at the he last traces of soap slowly disappear down the drain.

Things happen for a reason. I just need to let this new scenario play out. Yep, I've really, really been quite fortunate. Not many people are blessed with two wonderful daughters like I have, or have the good fortune to spend time with someone like Sandy. God, I miss her.

He closed his eyes and let the spray of the water beat against his face.

Man, this shower feels good. The three-hour walk I took and the battle with my filthy pickup nearly did me in. Yep, I loved Sandy without reservation for thirty-five years, more if I count back to when we first met. Yeah, yeah, I loved Trish, too, but now it's almost like she's intruding.

Suddenly an almost forgotten identity invaded, roaming about unchecked. He no longer felt uneasy or experienced uncomfortable, churning feelings when he thought about Diana.

Elusive, mysterious and temperamental as hell, is the best way to describe her. Oops, I almost forgot. She was also fiery, fiery as hell. At the time, even though she was nothing more than an infatuation, she had a hell of an influence on everything I did.

He grinned, shaking his head.

Fuck, every girl I met had to pass the Bodia test. Whether I want to admit it or not, she was very influential.

He again let the spray from the nozzle hit his face.

Sandy was so much more than just the mother of our three kids. She was my partner, my best friend, and my lover. I loved her so much, and there won't be a day I won't miss her.

He cleared his throat.

She was, and probably will be, the most important entry in my diary even though I know it's possible to love more than one person.

He lowered his head.

Cancer's not a fair way to go. Sometimes I almost wish it had been me.

Suddenly his memory tapes started to rewind, stop, and then slowly begin to inch forward. It was March of 1962.

"Honey, I'm pregnant. I know the timing stinks, and it's not the first thing you wanted to hear when you got out of the service, but I guess it's impossible to orchestrate everything. Think we might have gotten a little carried away while you were home on leave?"

"We were going to get married in June, anyway. I guess getting out early so I could go back to school was a good idea, after all."

"Like you had doubts."

"So, what are we waiting for? Why don't we drive down to Reno and get married right now? Course, I suppose if you want a formal wedding, we could wait."

He recalled that at that precise moment, she smiled in the way she always did when she wanted to make him feel like clay in the hands of a potter.

"No, I think going to Reno is a terrific idea. I'll take the week off so we have plenty of time for a honeymoon. I've never been big on formal weddings, anyway."

Inching forward, the tape stopped. It was June of 1964.

The doctor was sitting next to him in the waiting room. He grimly revealed that it was a boy. After placing a hand on his knee, the doctor went on to say that there was nothing he or his staff could do.

The baby just didn't have enough time to develop.

Smiling reassuringly, the doctor told him that Sandy was fine but reminded him that she would need some time to recover even though she was strong physically. He was more concerned about the psychological impact, instructing Lee to be strong for Sandy.

"When can I see her?"

"I don't see why you can't see her right now. She's waiting for you."

When he entered Sandy's room, he went over to the bed where she was lying, bent forward, and kissed her on the cheek.

"I'm so sorry. The doctor told me it was a boy."

Choking back tears, she nodded.

"I want to give him a name and have a proper burial for him. God, I wanted the baby so bad."

His eyes filled with tears.

"I know, so did I. You're so beautiful. I love you more than I can possibly describe."

"And I love you. Do you still like the name, Addison Thurman?"

"Not a very good way to carry on the family name, is it?"

"We'll have another baby. I want to try again, real soon. Please hold me?"

Speeding ahead, the tape slowed, then stopped. It was February of 1975 and the nurse had just entered the waiting room.

"It's a girl, Mr. Grady. You have a beautiful eight-pound, four-ounce girl. She's perfect."

"Thank God. We've waited so long for this moment. Are you sure she has ten fingers and ten toes?"

"Positive! Ah, your wife is waiting to see you. She wants to name her Nicole Marie."

"Thanks, nurse. You've made my day. Hum, I thought we had settled on Marri Nicole. Oh, well, Nicole Marie's a great name."

The tape raced forward to December of 1977 and again a nurse entered the waiting room.

"It's a girl, Mr. Grady. You have a beautiful eight-pound, ten-ounce girl. She's perfect."

"Do you remember the drill?"

"I sure do."

"Ah, she has ten fingers and ten toes."

"Did you know that more than anything I wanted another girl?"

"Wonderful, you got your wish."

"Is my wife able to see me?"

"She wants you to hurry. She's holding your new daughter. By the way, she wants to name her Charon Amanda."

"Hum, here we go again. I thought we had settled on Mandi Charon. Charon Amanda, hum, great name."

It was April 1997, and he was sitting across the table from Sandy. He searched her eyes for a clue about the mysteriously arranged dinner date.

"Do you remember, Lee?"

"Of course! This is where I proposed to you. Can you believe thirty-three years have passed already?

Her eyes misted.

"Seems like only yesterday. The years we've spent together have been extraordinary, filled with so much love. I couldn't have asked for more. It's all been like a fairy tale."

"Yeah, and you're my fairy princess."

He reached across the table, taking her hand.

"So what's up? It's not our anniversary or one of our birthdays. This isn't even the anniversary of when I proposed. Are we celebrating another successful prosecution?"

Sandy squeezed his hand, tears filling her eyes.

"The time has come when I have to ask you to be brave and strong for the girls, Lee."

The color drained from his face, fear suddenly gripping his heart.

"You're a lousy liar. The doctor had bad news, didn't he?"

"I have six months."

The tape slowed, finally stopping. Slowly it started to creep forward. Sandy was lying in the bed, a peaceful smile consuming her expression.

"I've loved you since I first met you at the Erb Memorial in Eugene during the state basketball tournament."

"I fell in love with you there, too, remember?"

She smiled weakly.

"I suspected as much. You never were good at hiding your feelings. You know you never had a chance, don't you?"

"I was willing prey."

"Humph, think again. You were anything but easy. You forced me to plot every imaginable way to trap you. I thought you'd never make an honest woman of me."

"You didn't have to work very hard, Sandy. I wanted to spend my life with you the moment I laid eyes on you. Yeah, except for a few interruptions, things have worked out pretty well."

Tears flooded his eyes.

"Loving you was so easy. Our marriage was made in heaven."

She smiled faintly, her eyes suddenly seeming to be miles away.

"It was, wasn't it?"

She squeezed his hand. Suddenly as quickly as she had retreated into her private shell, she returned.

"I have three requests, Lee."

"They're yours. Just name them."

"I want you to be brave for the girls. Do your grieving in private. They will need your strength."

Tears streamed down his cheeks.

"You hit me with a tough one, there, but, ah, I'll do my best."

"I know you will."

She squeezed his hand again.

"I want you to be my eyes. Watch out for them. Keep them safe so they can have the same wonderful life we've been able to share."

"That will be easy to do. You know I love them more than all the rolled oats in the world."

"You and you're rolled oats."

She paused for a moment searching for the right words. Finally squeezing his hand with a sense of finality, she again met his gaze. Finally, she continued as though delivering a final summation during a trial.

"I want you to marry again, Lee. You're too young to spend the rest of your life alone."

"Not fair. You know nobody can replace you. You're my life. There's no way I could be happy with anyone else. Besides, what will the girls think?"

A smirk suddenly consumed her expression. By now familiar, it was a look which indicated what she was saying should be understood because of its simple clarity and perfect logic.

"The girl's will be relieved. They know what a big baby you are, and how you hate to be alone."

She smiled weakly.

"If I know them, they'll start a dating service to find you a companion. I don't know, Sandy. I don't see myself running around looking for someone. It just doesn't seem right."

She shook her head for emphasis.

"It looks like you've forced my hand. Um, I've known something for several years I've never told you about, something I've been forced to keep a secret."

"Really! We've never kept anything from each other, before. What could be so important you'd keep it from me, anyway?"

"Call it a gift, an extraordinary gift."

"I like a mystery, but right now I'm not sure I'm in the mood. So, what's this secret?"

"Patience, Lee, the time's not right, just yet. You'll have to wait a little while longer, but when the time comes, I think my little secret will make you very happy. Just promise me you'll keep an open mind. If you remember to count to three before you react, everything will be just fine."

She fixed her eyes on his. "Promise you'll not let your intolerant nature take over like it usually does."

"Okay, but who's going to share all of this happy news with me?"

She put her forefinger to her lips.

"Also, please try to understand when you find out why I had to keep the secret from you for all of these years, that I never wanted to intentionally deceive you. Circumstances forced all of this on me."

He looked into her eyes, studying her carefully. Tears streamed down both of their cheeks.

"There will never be another person like you, Sandy. I love you more than life itself."

"Thank you for sharing your life with me, Lee. You have made me the happiest women on the face of the Earth."

Not wanting to recall the painful memories of Sandy's funeral, an affair of finality, he started to briskly towel himself after turning off the shower and stepping carefully out of the shower onto the bath mat.

It'll sure be good to see Mel again. What's it been now, a month? He's probably been fishing with Byron. Sure is strange. When it's all said and done, Trish could turn out to the epilogue of the whole story. She may be the one who adds meaning to it all. At least, now, I know what Sandy meant about being able to have the son we weren't able to keep. At ten minutes before seven, Mel wheeled his late-model sedan into a thirty-minute parking space on Oak Street, a short block from the Benson Hotel. He turned off the ignition and turned towards the back seat.

"We've arrived. Sure this is close enough?"

"This is perfect."

Scanning the street, he nodded with approval.

"There she is. It looks like she just came out the door. I'll be back in a moment." Grunting as he exited the dark blue Mercedes, he again confirmed, "I'll only be a moment."

"Not nervous, are ya? What's there to be uptight about. No big deal, she's only just returned from the dead."

Lee knelt down so he could see both Mel and Lynne.

"Please, do me a favor. Stop speculating, she's just an old friend. And as far as all the talk about her returning from the dead goes, let's not go there. When I find out more of the details, you'll be the first to know, but for now, just know what I know, you know."

He rose, turning towards where Trish waited and started to close the distance between them.

Shit, I better get a move on before she gives up on me. Whoa, slow down a little. I don't want to come across as being too eager. Who am I kidding? I can't wait to see her. Shit, I must look like some comic from the silent movie era. I wish I really knew how I feel about this meeting. Part of me is thrilled to death, but another part of me is feeling a little betrayed.

On his right, two cars raced up Oak Street. Charging through a red light at Broadway Street, their impatience was obvious. Apparently nobody wanted to face the delay imposed by hitting another red traffic signal in the heavy early evening traffic.

Wow, look at her. She's still looks the same. Even after all these years, she's as elegant and beautiful as she always was. Yes, the years have been kind to her. Her figure's still trim, and, ah, all the curves are still in the right places. Yep, very appealing. I can't even detect a hint of gray in her hair. Remarkable!

At the corner, while waiting for the pedestrian signal to change, he continued to study her.

If she has changed, the transformation is barely perceptible. Nice tan! Hmm, she still doesn't need any assistance from the cosmetic industry. Fuck, she is still flat out beautiful, ah, looks like she did years ago.

The light changed. With some hesitation, he entered the crosswalk.

Good! She hasn't seen me yet. Maybe she just hasn't recognized me. I'll sneak up on her from behind, surprise her.

A moment later, inhaling the intoxication cast by her mild perfume, he prepared to launch the surprise attack.

"Pardon me, would you happen to have some spare change for an old sailor?"

Trish wheeled around with an expectant smile gracing her face.

"Oh, my God! It's so good to see you."

Unabashed, she thrust her arms around him.

"Shame on you. You're not old. You look great."

Responding awkwardly to her embrace, he argued, "Time changes things, Trish. It waits for nobody. Sometimes it can be so cruel, but in your case, it's been extremely kind. You look just like you did years ago. It's so good to see you. It's been too long."

"Yes, yes it has." Frowning, she added, "You're going to make me earn this, aren't you?"

She leaned forward and kissed him softly on the lips.

"Um, where are you parked?"

He took another appreciative survey, casually nodding toward the corner behind him.

"Down the street." Chuckling, he teased, "I didn't want to create a traffic jam in front of the hotel. Ah, how should I introduce you?"

"Like I said on the phone, you've only know me as Trish Knight. No sense in changing, now."

"Well, then, come on, and let me show you to the limousine. The night on the town has just begun."

He grinned, glancing at her as they entered the crosswalk.

"I'm hoping we have plenty of time to catch up on all of the years we've missed."

"The catching won't be difficult, it's the letting go that could cause us some trouble."

She squeezed his hand affectionately.

"Um, I don't have a curfew, so I see nothing preventing us from getting everything out in the open."

"Yeah, I'd like to make arrangements to meet my son."

"I think it can be arranged. Ah, have you explained everything to your friends?"

"Basically I just told 'em what was in the letter. Ah, here we are."

Smiling broadly, Mel rolled down the window.

"You can make the introductions after you get in the car. Curbside greetings in Portland are a little risky."

After seating themselves in the back of Mel's car, Lee chuckled nervously.

"Um, Trish, I would like to introduce you to Mel Marsters and his wife Lynne. In case you haven't guessed, we go back a few." Adjusting his position, he continued, "Um, Mel and Lynne, I've told you about Trish Knight, the old friend from my Navy days in Newport. Well, here she is, in the flesh."

"Shame on you, Lee! I may be a friend from the deep past, but I'm certainly not old. Ancient, maybe, but never old."

"Good for you, Trish. We women have to remind these guys every once in awhile, old is strictly in the eyes of the beholder. I don't think you're old, either. If you're old, how is everyone going to describe the rest of us?"

Mel chuckled, his eyes twinkling. Lee could tell that he was about to unveil another humorous remark.

"Just like two women, they always stick together when it comes to the important things in life like discussing age or dress sizes. You know, Trish, being called ancient isn't so bad. I had an old tractor once. It always seemed to run better when I told it, for an ancient son of a gun she was still near and dear to my heart."

"Hmm, this is the first time anyone has compared me to a tractor. Maybe a Dear, but never a Deere."

"Touché! So, do you run like one?"

Relentless to the end, Trish fired a final volley, "Dear, Deer, Deere, what confusion has been thrown in our laps by the simple addition or interchange of a few vowels. Oh, Mel, why don't you be a Dear and not think of me as a Deer, especially not a Dear Deere?"

"I like your friend, Lee. She's sharp as a tack."

Lee chuckled at the exchange. Mel's humor was still razor sharp, but Trish could more than hold her own.

"Okay, stop already. Have you two figured out the difference between a deer and a Deere?"

"Yes, dear. Ah, Trish, we were thinking about going to an establishment known for its tasty steak dinners. How does a good steak dinner strike you?"

"Time hasn't dulled my appreciation for a good steak, Mel. I may not run like a deer but I can still chase a good cut of meat."

"Touché again! Lee, it looks like you and I had best stay out of the way. Mel and Trish seem to be getting warmed up for a long tennis match."

The Ringside's service, outstanding, enhanced a delicious meal. A predinner entrée of the Ringside's famous onion rings awakened the voracity of their appetites, the dinner salads garnished them, but the mouth-watering steaks truly mesmerized their palates. Enough should have been enough, but, no, they decided to risk their waistline measurements on decadent chocolate cake and snifters of cognac.

After forcing down the last bite of his cake, Lee moaned, "I'll burst if I eat or drink anything else."

Mel chuckled, rubbing his stomach.

"I'm afraid someone's going to have to wheel me out of here."

"Oh, no you don't. I don't see a hand truck around, and I'm sure as hell not going to try packing you. You know, Trish, this is the first time I've ever seen those two beg for mercy at the dining table?"

Straight-faced and completely serious, Trish waved her handkerchief in mock surrender.

"If only we hadn't ordered the onion rings."

"Sorry, Trish, but I think we all went way past the onion rings. Ah, your stories about Boston have been very interesting." Staring into the distance, Lynn mused, "Someday, someday, I'm going to talk Mel into taking me back there."

Mel looked at his watch, nudging Lynne. "I think it's time for us to start heading out. We still have an hour's drive down the freeway. Don't forget, I have to get up early tomorrow morning."

"Gonna try your luck out on the big pond, Mel?"

"Yeah, I thought I'd try to hook into one of those Chinooks everyone seems to be landing lately. Want to go? We have plenty of room."

"I think Trish and I have a lot of catching up to do." Winking at Trish, he added, "I think tonight could be a late one."

"When friends are reunited, there's always a lot to catch up on. Ah, Mel, are you trying to say it's nearing your bed time?"

"Heavens to Betsy, Trish, Mel's bedtime has already come and gone. Normally, he's already in bed by this time."

As they got up from their table and started to leave, Lee turned to Mel, suggesting, "Why don't you drop us off at my place?"

Trish slipped her arm through his, squeezing it affectionately.

"Good idea. So far, we've barely scratched the surface. Trying to explain the counterfeit accident report might be my biggest challenge."

Frowning, Lynne probed, "I must have missing something."

Mel nudged her, shaking is head.

"Never mind, it's none of our business." Turning to Trish, he added, "Good luck, I'm dying to hear the finer details myself. No pun intended. Um, how long do you plan to stay in Portland?"

"I was thinking of leaving tomorrow, but with things developing the way they are, I may wait until Sunday evening before I leave. My business meeting in Seattle isn't until Monday morning."

"Do you have plans to return to Portland any time soon?"

"Could be. Actually, I'm trying to buy some property in Lake Oswego."

Lee shifted uneasily in his seat.

"So are you going to take me up on my offer and go to the football game next Saturday, Mel?"

"Yeah, I don't see why not. Isn't next Saturday the twenty-third?"

"Yeah, why do you ask?"

"Oh come on, Lee. You know all of Mel activities revolve around fishing. Ah, he has a trip planned for the twenty-fourth. He just wants to make sure nothing interferes. I swear, if there wasn't some damned fish to catch, Mel wouldn't know what to do with himself."

After a quick trip across town, Mel pulled his car into Lee's driveway, stopped and slipped it into park. He turned and looked at his friend.

"I'm going out for a two-day trip on a friend's charter trip in a couple of weeks or so to try my luck at catching some tuna. If you'd like to come, you're welcome. The price is right."

"I'll let you know. Right now I can't think of a reason why I can't."

He nodded, opening the door.

"Yeah, I'll let you know. Ah, thanks for everything. The evening was spectacular as always."

As Trish eased out the door, she patted the back of the front seat.

"I enjoyed myself immensely, Mel and Lynne. It was such a pleasure to meet you. I hope we'll be able to get together again, real soon."

Lee bent down as his friend rolled down the window.

"Give me a call and let me know about your fishing trip."

"I will." Grinning, Mel petitioned, "I'd also like for you to fill me in about what happened after Trish fills you in." Turning towards Trish, he added, "Thank you for joining us. Lee needs a chaperone sometimes to keep him in line, but you can always count on me if you're ever in need of any help. Ah, Lee, the next time your friend comes to town, bring her down to Salem. We'll see if we can't find some mischief to get into."

"Count on it."

After Mel backed out of the driveway, Lee turned to Trish, gently slipping his arm around her waist.

"I'd say let's go in the house so I can show you my etchings, but we're both well past playing those kind of games."

"Oh really? I think it might be fun to see your etchings."

He grinned nodding thoughtfully.

"I'm not sure I remember the drill."

"Humph, think you've forgotten, what about me? After thirty-some years, I've come to understand how a nun feels."

She waited for him to enter the code into the garage electronic pad.

"While you're giving me the tour, I'll try to fill in some details not covered in Sandy's letter."

He nodded, waiting for the door to open.

"Only special people enter my home this way."

He took her hand and gently squeezed it.

"You know when you're more than just an acquaintance when you're allowed to enter a person's home through their utility room."

"I hope it's neat and tidy, or maybe you have a maid."

"No such luck. Ah, what will it be, a quick tour or should we stand here and do the laundry together?"

"It's not like you to make small talk, Lee."

She smiled, entering the kitchen.

"Lead the way. I may learn something about you by seeing your home. I've never see this side of you, before."

"Like I had the chance."

"Don't reminded me."

As she followed him from room to room, she shook her head in awe.

"My God, Lee, Sandy's presence is everywhere! I almost feel like she's looking over my shoulder."

"If she is, she has a smile on her face. Ah, just how well did you know her?"

"Really well. When you work with someone as long as we did, under the circumstances we had to.... Ah, well, you get to know them very, very well."

"Uh-huh. Well anyway, back to your comment about her inhabiting this place. Ah, she has only been gone for a year, and, ah, I haven't exactly been in a rush to get rid of her."

"Understandable. Gosh, every nook and cranny reflects her influence. My God, Lee, its so eerie. No matter what room I go into, it's almost like she's still here."

Trish studied a photograph.

"Sandy was even more beautiful than I remembered."

"Yeah, there was a spiritual beauty as well."

His eyes started to glaze over.

"I loved her so much, and there's not a day goes by I don't think about her. She was very special."

"Indeed she was. She made me feel so comfortable during the investigation."

She continued browsing about.

"Sandy did an excellent job decorating your home."

"This was her sanctuary, where we raised our two daughters. She poured her heart and soul into this place. Ah, if you need to use the bathroom, it's on your right."

"Thank you. The drinks we had at the restaurant are starting to catch up with me."

A few moments later Trish entered the living room.

"Good, I see you poured the wine."

"I hope Riesling is okay."

"My favorite. Um, you aren't trying to ply me with spirits, are you?"

"Would it do me any good?"

"You'll never know if you don't try."

Pensively she took a sip from her glass of wine.

"This is excellent."

"As I recall, you usually drank beer."

"I've broadened my horizons."

She allowed her eyes to meet his steady gaze.

"Wow, where to begin? Where should I start?"

"I think Sandy covered things pretty well. All you need to do is clarify a few points."

"It might be easier if you tell me what you want to know."

"How about starting with your trip to New York to see Diedra, ah, just before you were supposed to meet me in Milwaukee?"

"Just something my father put in the letter to make the story about my death seem more plausible. Um, Diedra agreed to play along, ah, right from the start."

"Okay, so, was getting your driver's license also a hoax?"

"No, I got my license, but I didn't get it until I was in the witness protection program." Grinning, she clarified, "For identification purposes."

"Did you really wanted to get married as you said?"

"Absolutely. If it hadn't been for my father, I would have married you in a heartbeat."

She exhaled violently, causing her lips to flutter and making the sound most people don't want to emit in public.

"The week before your graduation from boot camp, my father informed me I had to make a decision. I could have an abortion, or I could keep the baby and marry Vinnie Trabago. In his mind, marrying you wasn't an option."

"What if you had come to Milwaukee as we planned?"

"You would have been swimming with the fishes."

"Meaning?"

"A common Sicilian term. He would have had you eliminated, killed. If you knew my father, you would know there was absolutely no point in arguing with him. As far as he was concerned, the matter had been decided."

"Why did he want you to have an abortion? I thought he was a Catholic?"

"He was. Ah, well, first of all, there was no way he was gonna allow me to marry you. He told me he wasn't gonna stand by and watch me bring a bastard into the world."

"Why wouldn't he have let you marry me? He knew nothing about me."

"Who knows? He never said. Ah, the only explanation is that he was Sicilian, very traditional. In his mind, I suppose, he felt I should marry an Italian. Then there was the matter of trust. He knew Vinnie would keep me in line. Yeah, good old Vinnie Trabago, what an asshole."

Tears filled her eyes.

"You'll have to believe what I told you before. We were married in name only. I never shared his bed."

"I remember calling Diedra after reporting to my first duty station in New York. She mentioned Patti Anatolia served as her maid of honor."

"Did you suspect it was me?"

"Something didn't ring true, but, no, I can honestly say I was completely unaware. Looking back, all the clues were right under my nose."

"I guess my father did his job. Well, anyway, after Diedra's wedding, we moved to Seattle. My father decided to run the family business from there and let Vinnie run the East Coast operation from Boston."

"But you say you moved to Seattle. Didn't you live with him?"

"I moved back to Boston after things cooled down and we found out you had been assigned to a ship in the Brooklyn Naval Yard."

"If you knew where I was, why didn't you try to contact me?"

"My father and Vinnie threatened to take my son away from me. I also wanted to prevent you from having an accident, if you know what I mean."

"So when and how did you come to meet Sandy?"

"When things starting getting totally unbearable, I got in contact with an old college friend of mine who worked for the police force in Boston. The next thing I knew, Sandy and I were discussing turning state's evidence and getting me into the witness protection program."

"You met with her in Boston?"

"Yeah, after you got back from the Barents Sea, she and some of the Feds who worked with the witness protection program met with me."

"How did you come up with the name, Pat Brown?"

"Easy. I graduated from Brown University and my first name was Patricia."

"Clever. And where did you live when you went into the witness protection program?"

"Montana. My son and I lived in Missoula, Montana."

"How long?"

"Let me see. My father and Vinnie were indited in February of 1964. Um, we lived there from 1964 until about a year ago when I buried the hatchet with my mother. I'll let you do the math. Um, my son moved to Seattle in 1993 after graduating from the University of Montana. He's an accountant."

"So, have you told him about me?"

"I told him everything when my mother and I reconciled our differences. By the way, his name is Truman."

"Is he gonna want to see me?"

"Of course. He wants to meet you very badly. In fact, he and his wife, Leeanne, are in town with me. They're also staying at the Benson."

"When? When can I meet him?"

"Tonight?"

"Yeah, tonight would be fine. Ah, first, though, I have a few more questions."

"Fire away. This is going faster than I thought. It's also a lot easier than I thought it would be."

"It is, isn't it?" Well, anyway, after your father and Vinnie were put away, why didn't you try to contact me?"

"You were happily married. There was no way I would have ever done anything to mess up your life. Besides, I had too much respect for your wife. In so many ways, she saved me and my son's life."

"Until you turned state's evidence, you lived in Boston with Vinnie."

"Yeah, and every weekend he went up to his house in Connecticut. He was fucking some bimbo from New York."

"I know. Her name was Margaret Gilland. Special people in her life called her Peggy. Ah, she and I saw each other." Chuckling, he confessed, "We spent some time together when she wasn't fucking Vinnie."

"Um, what did you call her?"

His face reddened.

"Ah, Peggy. Ah, she told me about Vinnie. I guess she really loved him."

"No, she loved his cock."

"Small world, huh?"

"No kidding. Were you doing her, too?"

"We had our moments."

"Yeah! Remember when you called, right after you got the letter letting you know about my demise?"

"Yeah."

"My dad almost shit his pants. I know because I was there and overheard everything. At least I heard my father's side of the conversation."

"It seemed kinda strange. There was a lot of yelling and screaming going on in the background. Your father was real uncomfortable, very distant. Not a grieving sort of distant, just distant."

"The yelling and screaming in the background was Vinny beating the shit out of me because I was trying to get to a phone. When you hung up, he told me if ever I tried to contact you again, he'd have Vinny whack you. I just sort of gave up after your call. I didn't want you to get hurt. For what it's worth, you were the first, the last, and my only lover."

"When did you find out Sandy and I were involved?"

"Just before you got out of the service. I told her if she ever hoped to get me to testify she'd have to keep her involvement with me a secret. I made her swear never to tell you about me. Before Sandy died, she contacted me and convinced me to contact you and let you know the truth. She mentioned something about giving you back the son you lost years ago."

"Yeah, and now here we are."

"I still love you. I know it's probably too late for us, but you still have time to forge a relationship with you son."

"Yes, yes I do. Ah, for whatever it's worth, I was in love with you. I would have married you, you know."

"I know."

"I loved you then, and I guess I still do. I don't think loving you diminishes my feelings for Sandy."

Tears filled his eyes.

"Cancer is such a humiliating disease. When it makes up its mind to do its work, a date is set with the grim reaper. Then, like an old fashioned family doctor, it makes a house call."

He paused to wipe his eyes.

"Throughout it all, she was so brave and upbeat. The day she died, I held her and told her I would always love her and I would always miss her. You know what she said to me?"

"No, what?"

"She told me if she had it to do all over again, she wouldn't change a thing. She asked me to look after the girls and not be afraid to find someone else. Ah, the last thing she told me was, we had an appointment in heaven. A minute later she took a deep breath and went to sleep."

Trish put her arms around him, whispering softly in his ear as she tenderly kissed his cheek, "It's okay. Let it all out."

"I thought I had cried all the tears I had. Ahem, enough of this. I have a son to meet. Yeah, I'd like to meet Truman."

Trish placed her hand on his and then closed her eyes.

"The time's right, Lee. He wants to meet the father he never was allowed to know."

She hesitated, looking away.

"He's not expecting much."

"Well, what are we waiting for?"

"It will mean a lot to him. Like I said, he's hopeful."

"Hopeful?"

"Yes, I think he really hopes you will want to form a relationship with him. He would also like to meet his sisters."

He nodded meeting Trish's gaze, trying to communicate what they both were feeling.

"What about us? Is there any hope or is it too late?"

"Ah, I think I should call Truman to see if he and his wife are ready for a visitor."

She smiled, checking her watch.

"It's eleven. I don't think it's too late, do you?"

He shrugged, taking another sip of wine.

"Call him. For a meeting like this, it's never to late."

"If he agrees, what time should I tell him we'll be there?"

"It takes about thirty minutes to get there. How about an hour?"

"Why the delay?"

"Ah, you didn't answer my question. I'm thinking maybe I'll have to work a bit to pry one out of you."

Her eyes twinkled as she reached for the portable telephone located on the coffee table in front of them.

"That's fair. Give me a minute. As soon as I call Truman, I'll give you an answer."

A moment later, she laid the portable back on the coffee table.

Searching her eyes for a hint, he inquired, "Well, was he surprised?"

"No, just a little nervous. They'll wait up for us. Now I suppose you're waiting for my answer."

"You may not be prepared to give me an answer right now."

"Oh, I'm ready alright."

She paused. Slowly she allowed her eyes to meet his steady, expectant gaze.

"I've never stopped loving you, Lee. If you want to take another chance to see if we can do it right this time, I'm all in favor. Ah, I have every intention of kissing you."

He greeted her movement towards him, meeting her lips responsively.

After a guarded but passionate exchange their lips parted.

"When I asked you to join me for dinner this evening I was afraid of being disloyal to my wife's memory."

"And now, what about now?"

"Something inside of me keeps tugging at my heart, hoping you and I will be able to rekindle the magic. I know it will take time and patience, but I'm willing to try if you are. I still love you, Trish."

"The magic's still there, isn't it? I love you, too. I'm in no hurry. At our age, rushing into things isn't good for the heart."

He bent forward and kissed her again. After their lips had parted, he chuckled.

"The shoe seems to be on the other foot this time, though."

"What do you mean?"

"This time, you're waiting for me."

"Yeah, but nobody's father's calling the shots."

A few minutes later, Lee backed his Honda out of the garage, and turned towards downtown Portland. He glanced towards her.

"With your father's business sold, what are you gonna do now?"

"I really don't need to work anymore, but I'm going to do some design work for a company called T-Line. It specializes in women's sport's apparel."

"And Truman? Will he continue on as an accountant?"

"Probably. He really likes Seattle."

"And you?"

"Drive slowly, Lee. We still have plenty of time. Truman has waited this long, another few minutes won't hurt."

She looked out the window at the Willamette River.

"Like I said earlier, I've been looking at property in Lake Oswego."

"Don't you like living in Seattle?"

"It's okay."

She cuddled close to him as they pulled up to a stoplight.

"With a little encouragement, I could get used to the Portland area real quick."

He leaned over and kissed her. The response was immediate, passionate. A moment later their lips parted.

"Would it help if I asked you to stay?"

She grinned, gently pushing his away.

"The light changed. If you don't hurry up, we'll get arrested for loitering."

"It would be nice if you decided to move to Portland."

"Maybe you'd like to see the home I've been looking at."

Again she stared out the window.

"We had our moment years ago, not to say we still don't have a few left, but maybe what time we have left should be spent working on being good grandparents. It's going to happen someday, you know."

"I know. Um, Nicole and Jason have been married for over a year now, and Charon, well I still have to walk her down the aisle."

"Truman and his wife have been married for over a year, too. They don't say much, but I think they're trying real hard to get pregnant."

She turned towards him, smiling with an inquisitive expression etched on her face.

"Honey, were you serious when you said you wanted to give it another go?"

Looking up, he noted the sign for the Hawthorne Bridge.

"There's the sign for the Hawthorne Bridge?"

"What about it?"

"I'd guess we have from ten to fifteen minutes before we knock on Truman's door."

"So…?"

"Our first go-round, relatively speaking took about fifteen minutes. We weren't doing too badly until your father decided to interrupt. This time, there's nobody to interfere. We just might make it work."

He glanced over at her, smiling.

"It's my way of telling you I want to give it try."

"You have the sneakiest way of saying yes."

"I wasn't being sneaky when I told you I still love you."

Ten minutes later, Trish knocked on room 712. A tall, handsome dark haired, man looking to be in his middle thirties answered the door. Trish smiled.

"Are you two decent?"

Truman nodded eagerly.

"Come on in. Leeanne and I have been expecting you."

His wife stood in the softly lighted room, soft music playing in the background.

"Mom, why don't you do the honors?"

"My pleasure. Ah, Leeanne and Truman, I want you to meet Lee Grady. Ah, Lee, this is my son, Truman, and his wife, Leeanne."

Stepping forward, Truman nodded, extending his hand.

"I'm pleased to meet you, Dad. I've heard a lot about you."

"All good, I hope."

Lee gripped his son's hand.

"I'm so glad to meet you. This meeting has been long overdue."

Lee smiled, moving towards Leeanne to give her a hug.

"And I'm please to meet you too, Leeanne."

"Ah, I'll turn the music off. Truman, would you mind pouring a glass of wine for us?"

"Good idea. I think I need a little reinforcement. I don't know about the rest of you, but I'm not sure I know exactly what to say."

"Sometimes words aren't necessary, Truman. If you're game, I'd like for us to get to know each other."

Truman smiled.

"Yeah, I'd like that. Yeah, I'd like that very much."

Lee nodded, looking at Trish.

"Your mother and I have a lot of catching up to do, as well. Ah, do you play golf?"

"A little. I just took up the game."

"Good, maybe you and I can go out and hit it around, sometime."